RAFAEL CORDERO

THE DESIGN ANALYSIS HANDBOOK

THE DESIGN ANALYSIS HANDBOOK

A Practical Guide to Design Validation

Revised Edition

N. Edward Walker

🔲 NEWNES

Boston • Oxford • Johannesburg • Melbourne • New Delhi • Singapore

Newnes is an imprint of Butterworth–Heinemann.

 A member of the Reed Elsevier group

 Recognizing the importance of preserving what has been written, Butterworth–Heinemann prints its books on acid-free paper whenever possible.

 Butterworth–Heinemann supports the efforts of American Forests and the Global ReLeaf program in its campaign for the betterment of trees, forests, and our environment.

Library of Congress Cataloging-in-Publication Data
ISBN 0-7506-9088-7

British Library Cataloguing-in-Publication Data
A catalogue record for this book is available from the British Library.

The publisher offers special discounts on bulk orders of this book.
For information, please contact:
Manager of Special Sales
Butterworth–Heinemann
225 Wildwood Avenue
Woburn, MA 01801-2041
Tel: 781-904-2500
Fax: 781-904-2620

For information on all Butterworth–Heinemann publications available, contact our
World Wide Web home page at: http://www.bh.com

10 9 8 7 6 5 4 3 2 1

Printed in the United States of America

CONTENTS

Contents

FIGURES AND TABLES

Figures and Tables

EQUATIONS AND EXPRESSIONS

Equations and Expressions

PREFACE

This Handbook has several goals. The first is to help project engineers and reliability/quality managers sleep better at night, by providing a straightforward, no-nonsense guide to practical design validation. A second goal is to help design engineers who may be neglecting their math skills to recapture the wonderful sense of satisfaction that results from truly understanding a design, an understanding which leads to design excellence. And finally, this Handbook is intended to be used as a practical design assistant; numerous practical tips and insights are provided which can be used to avoid a wide variety of design hazards.

Although the subject matter is serious, the Handbook has been prepared in a light and hopefully entertaining manner, based upon my belief that professionalism and stuffiness need not be synonymous. Also, it is hoped that the Handbook will serve as a bridge between the disciplines of design and quality/reliability. It has been my observation that although these fields should mutually support each other, the practitioners tend to segment themselves into their own niches. In fact, the methodologies and examples proffered by those in the quality arena to help design engineers are sometimes embarrassingly impractical or even wrong, leading designers to shrug off or even laugh at topics which otherwise may be of great benefit. The key is to meld the design engineering process with quality concerns, which is the perspective from which this Handbook was written. The straightforward design validation techniques described herein are the result of several years of "real world" design engineering, augmented by the practical application of design optimization and probability analysis techniques.

The Handbook emphasizes the essential role that analysis plays in creating reliable, cost effective, optimized designs. By analysis, I do not mean the manipulation of complicated mathematics; tedious calculations of matrix equations or other complex constructions generally do little to illuminate a design. No, analysis is used herein in its classical definition of *to think*, to understand. Generally, the related math embodies basic equations from which it is possible for the designer to envision the boundaries of performance, and to identify those factors which have the most influence. The equations presented in the Handbook will be easily grasped and utilized by those with an engineering background, and for the most part can be solved with a spreadsheet, basic math-based software, or other simple calculation aids.

The underlying model used to exemplify the discipline of practical analysis is the methodology called *Worst Case Analysis*; WCA is an excellent design validation tool which requires thoroughness and analytical thinking. An advanced version of WCA will be described herein, based upon the proven Worst Case Analysis *Plus* (WCA+™) methodology developed by Design/Analysis Consultants, Inc. In addition to identifying the limits of performance, advanced WCA generates sensitivities and probabilities which can be used for design optimization and risk assessment. It is shown herein how these latter topics, generally associated with quality/reliability functions, can be integrated into the design analysis process and implemented by the design engineer in a practical manner.

CHAPTER 1, "An Introduction to Design Analysis," describes how analysis is applied to the overall product quality process. The important analysis tool called Worst Case Analysis (WCA) is introduced. WCA definitions, misconceptions, cost effectiveness, and case histories are provided to illustrate the usage and benefits of WCA.

CHAPTER 2, "How To Perform A Worst Case Analysis," defines the ten key analysis steps. The chapter describes how to identify worst case solutions, calculate the probability of occurrence of an out-of-specification (ALERT) result, perform a risk assessment, and use the analysis data to correct and optimize the design.

CHAPTER 3, "Design Validation Topics and Tips," explains how to screen an analysis to save time, reviews computer-aided-analysis tools, describes how to ensure realistic results, and discusses the topics of nonlinear equations, software validation, tolerances, and accuracy.

CHAPTER 4, "Safety Analyses," describes and reviews the supplementary analyses used for critical safety-related designs: the Single Point Failure Analysis, the Fault Tree Analysis, and the Sneak Path Analysis.

CHAPTER 5, "Bad Science and Other Hazards," describes design engineering pitfalls which can spring from unexpected sources.

CHAPTER 6, "Electronics Analysis Tips and Equations," provides a sampling of some of the most common yet misapplied, pesky, or in general tough-to-do-right electronic circuits. This chapter contains useful design tips and equations and numerous examples.

Appendix A, "How To Survive An Engineering Project," presents essential advice on coping with high-tension development projects. Appendix B contains DACI's Worst Case Analysis Standard. A sample Worst Case Analysis report is provided in Appendix C.

Some of the material contained herein was obtained from DACI newsletters and application notes published earlier. However, most of the material is new, and was derived from DACI's extensive experience in the provision of Worst Case Analysis and related technical services. Although most of the examples are based upon electronics technology, the underlying methodology is applicable to all engineering disciplines, and can be readily applied to mechanics, optics, heat transfer, etc.

I deeply appreciate the support and advice I have received over the years from the many design engineers and managers with whom I have worked to advance the cause of rational and effective design validation. Special thanks are owed to Tom Campbell at Group Technologies, Ed Whittaker at the Honeywell LSIC Design Center, Doug Tennant at Smiths Industries, Charlie Leonard at Boeing, and Bob Pease at National Semiconductor. I also am indebted to my associate Diane Brett for her advice and help in the preparation of this Handbook.

Ed Walker
Tampa, Florida
November 1994

CHAPTER ONE ─────────────────

AN INTRODUCTION TO DESIGN ANALYSIS

Three Tales of Terror...

After its introduction two years ago, the latest XYZ model has exceeded all expectations, with sales and profits increasing rapidly each month. But this month, after receiving dozens of returned shipments and hundreds of calls on the warranty hot line, you know a disaster is brewing...

You're presenting your prototype circuit to a group from upper management, who heard that your department (and you in particular) had made a technological breakthrough. As the company president presses a control on the prototype, a crackling sound is heard, followed by a puff of smoke and a flame which singes his hand...

You received hearty praise from management after landing that large production contract. But here it is, nine months after contract award, and you've yet to ship the first unit; they're stacked against the walls, defying every attempt to get them to meet spec. Your job is on the line...

The villain in these vignettes is bad design, or more specifically, failure to identify a bad design until it's too late. This book will show you how to catch devastating design bugs before they wreak havoc on your company's profits and reputation. To start, let's consider the essential role that *analysis* plays in design validation as compared to *simulation*.

OUR FAVORITE EXCUSES FOR WHY
THE DESIGN CAN'T BE AT FAULT:
(OR "GEE, MAYBE I SHOULD HAVE DONE AN ANALYSIS!")

"The prototype worked perfectly."
"The computer simulations didn't identify any problems."
"Each unit was thoroughly tested."
"We've built it for years and until now never experienced any problems."

A Profession of Tweakers?

Do many engineers rely too heavily on simulation tools? Is part of our profession becoming a group of Tweak and Tune hobbyists rather than serious designers? Since simulation tools tend to encourage a cut and try methodology at the expense of a good mathematical understanding of circuit behavior, we fear that the answer is yes. Symptoms of this decline in engineering capability are a loss of basic mathematical capability, a reluctance or incapacity to employ analytical thinking, and an over-reliance on simulations, coupled with neglect of proper testing:

SIMULATIONS are *NOT* A SUBSTITUTE
FOR ANALYSIS and TEST

We note that this view is shared by others. In the plenary address at the 1990 IEEE Symposium on Circuits and Systems in New Orleans, Ronald A. Rohrer had this to say: "...And the SPICE generation? Many of our students merely hack away at the design. They guess at circuit values, run a simulation, and then guess at changes before they run the simulation again... and again... and again. And I strongly suspect that some of them who do this even escape to become "circuit designers" in industry..." (Mr. Rohrer is University Professor in the Department of Electrical & Computer Engineering and Director of the SRC-CMU CAD Center at Carnegie Mellon University.)

Yet another notable comment regarding over-reliance on simulation for design validation (from Robert Pease's "Pease Porridge" column in the 10 October 1991 *Electronic Design*): "...So, SPICE can be a useful tool in the hands of a good engineer, to help him analyze variations on the central theme. It's only when people start making euphoric statements that bread-boarding is passé and superfluous that we get worried. So, why do I keep saying these "terrible" things about SPICE? Well, it's not my fault that it only takes five sentences to say what SPICE is generally good for and five pages to list some of the pitfalls."

And finally, an excerpt from an editorial "Will the Real EEs Please Stand Up" by Ron Wilson in the 18 November 1991 *Electronic Engineering Times*: "It is the fundamental mathematics and analytical ability that distinguish an engineer from a technician. Our vaunted engineering schools are beginning, I believe, to give their customers very strong technicians' training under the label of an engineering degree. The result is not a shortage of people who can apply popular techniques to known problems. The result is a shortage of people who have learned to be real engineers. That is a crisis in the making."

Turning the Tide

This book is dedicated to helping restore the basic and essential engineering analysis discipline required for excellence. The underlying model used to exemplify this discipline is the methodology called Worst Case Analysis; WCA is an excellent design validation tool which requires thoroughness and analytical thinking. While acknowledging the key role that proper testing plays in design validation, this book will focus on the aspects of design *analysis* required for success.

Analysis as used herein does not mean the use of complicated mathematics. Too much complexity obscures the design interrelationships and provides camouflage for errors. No, analysis is used herein in its classical definition of *to think*, to understand. This requires the use of basic equations from which it is possible for the designer to envision the boundaries of performance, and to identify those factors which have the most influence on performance. Also, simplifying assumptions are often employed which enhance understanding while preserving design integrity.

Good analysis translates into innovative, reliable, and cost effective products; i.e., excellence. Math models allow the engineer to understand and expertly manipulate physical processes to create desired designs. Without a mathematical understanding of design parameters, how is the engineer going to ensure that a design is optimized or reliable? Without understanding, how can an engineer develop creative insights that will allow design refinements, enhancements, or even technology breakthroughs to be achieved?

A classically-trained engineer, relying on the principle that to engineer is to understand, would first research the literature. The engineer would next examine the available data, making use of existing equations and test data. If additional insights were required, the engineer would plot graphs, generate tables, or run special tests. New or expanded mathematical relationships would be derived, which the engineer would use to control and optimize the design. The engineer would then summarize this work, documenting the test and analysis results for future projects. To us, this process of reviewing, probing, *understanding* a design is real engineering.

ENGINEERING = UNDERSTANDING

1.1
AN OVERVIEW OF THE DESIGN VALIDATION PROCESS

The basic processes which determine the quality of a product or system are Design Quality, Component Quality, and Build Quality. Failure Analysis is used as a quality improvement mechanism for all three quality processes.

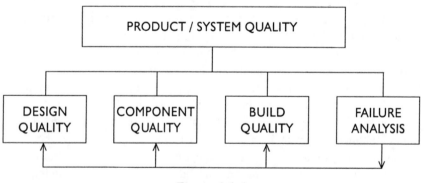

Figure 1.1-1
Quality Process Tree

Design Quality validates that a product or system will function as intended, assuming proper usage and no defects in workmanship or components. Design Quality also validates that a critical product or system will respond to the occurrence of a random component failure or operator error in a safe manner. Design Quality tools include design standards/guidelines (performance, testability, reliability, maintainability) and design validation.

Component Quality ensures that the components and subassemblies used in a product or system will meet their functional requirements over a defined minimum lifetime. Component Quality tools include vendor selection, component specifications, process controls, inspection, and test.

Build Quality ensures that manufactured assemblies are fabricated and assembled correctly and meet performance and compliance specifications. Build Quality tools include workmanship standards, fabrication specifications, training, inspection, and test.

All quality functions are (ideally) supported by a Failure Analysis team which investigates all reported anomalies, positively determines their cause, and validates corrective action implementation.

This Handbook focuses on the Analysis branch of the Design Validation portion of the Design Quality process:

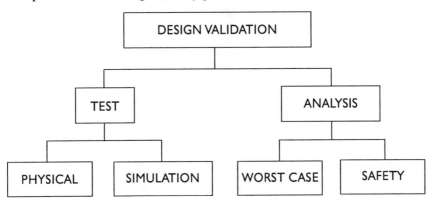

Figure 1.1-2
Design Validation Tree

Physical Testing employs one or more samples of a hardware prototype to demonstrate the basic functionality of the design. Physical testing is also used to identify weaknesses by exercising the design; e.g., thermal cycling, mechanical shock and vibration, etc.

Simulation is like physical testing, except a computer model is tested rather than a physical prototype. Like physical testing, simulation is used to validate basic functionality, but in addition can be used to gather more data more quickly than is generally the case with a prototype. For example, a matrix of simulations is usually performed on an integrated circuit design before the first circuit prototype is fabricated; e.g., "fast" and "slow" process parameters, high and low temperature, and minimum and maximum supply voltage.

Worst Case Analysis determines whether a system will meet its functional requirements, and whether any of the components in the system will be damaged or degraded. The analysis assumes proper equipment usage and no defects in workmanship or components. WCA is a cost effective means of screening a design to ensure with a high degree of confidence that potential defects are identified and eliminated prior to production and delivery.

Safety Analysis determines whether a critical system can exhibit unintended and unacceptable modes of behavior, and whether the system will fail in an acceptable manner following a random component failure or operator error. A safety analysis generally consists of one or more of the following: Single Point Failure Analysis (SPFA) (also called a Failure Modes and Effects Analysis (FMEA)), Fault Tree Analysis (FTA), and Sneak Path Analysis (SPA); see Chapter 4 for complete descriptions. (Note: Not to be confused with Safety Analysis is the safety compliance review, which addresses applicable safety standards; e.g., insulation, spacing, materials, etc. Compliance reviews are inspection and test functions performed as part of the Component Quality or Build Quality processes.)

The appropriate use of each design validation tool is summarized below:

VALIDATION TOOL	USE FOR	PURPOSE
1. Testing	All designs	Determine basic functionality and nominal bounds of performance; probe for weaknesses
2. Simulation	Many designs	Determine basic functionality and nominal bounds of performance; probe for weaknesses; support WCA

3. Worst Case Analysis	All designs	Identify limits of performance; risk assessment; design optimization
4. Safety: Failure Modes and Effects	All critical designs	Identify failure effects and assess their consequences
Sneak Path	All critical designs	Identify unintended modes of operation and assess their consequences
Fault Tree	Some critical designs	Determine the probability of occurrence of failure effects of particular interest

1.2
CLOSING THE GULF BETWEEN DESIGN AND RELIABILITY/QUALITY

The Quality department traditionally plays the role of setting quality and reliability standards and monitoring conformance to the standards. Quality engineers are also involved to varying degrees with problem investigations and corrective action implementation. These activities tend to be down-stream in the product development cycle, with design engineering being relatively isolated from many quality functions.

There has been much spirited debate over the last several years about integrating quality functions earlier into the design process, which is a good idea... in theory. In practice, this worthy goal is often thwarted by a gulf which exists between people who create designs (design engineers, project managers, technical directors) and people who try to ensure design integrity (reliability/quality engineers and managers). The former are busy trying to achieve functional, practical designs within budget and time constraints, and have little patience for the latter's attempts to inject new quality methods into the design cycle when such methods are based upon academic/theoretical data which demonstrate little to no understanding of the actual design process.

In many cases, although there may be some merit to the proffered assistance, the practical examples which would allow the designer to validate and readily employ a new technique are embarrassingly absent. In other cases the modest benefits of a new method are so overhyped as to create a reflexive backlash among designers, thereby greatly reducing its chance of acceptance. Talk of "six-sigma" quality, the "Taguchi" method of design optimization, or reliability "predictions" a la MIL-HDBK-217 often results in derisive snickers in the lunchrooms where designers congregate:

WHY RELIABILITY/QUALITY PROCEDURES ARE SOMETIMES INEFFECTIVE
The procedures are not validated by application at the detailed design level, and as a result are often embarrassingly impractical or even wrong.

The solution to this disconnect is to meld the design engineering process with quality concerns:

HOW TO IMPLEMENT EFFECTIVE RELIABILITY/QUALITY PROCEDURES
*Quality/reliability engineers **must** immerse themselves in the design process, and seek and use the inputs and detailed assistance of designers.*

Attempts to impose quality procedures without the requisite intellectual acceptance by the design team will fail. Academic blather and mumbo-jumbo will be defacto rejected, even if politely accepted at the superficial buzzword level. This is not insubordination, but just the predictable consequences of failing to provide the designers with the detailed, practical examples which would allow them to evaluate and use the new tools.

Sometimes the rejection of a new method is good, because the method is based upon bad science (see Chapter 5), and fully deserves a swift demise. In other cases, a valuable new tool is needlessly rejected simply because its merits are not explained to those who are expected to use it.

Therefore this Handbook has been prepared to bridge the design/quality gap, by providing the methodological foundation for WCA which will be appreciated by quality/reliability engineers, as well as specific practical examples which will demonstrate the value of Worst Case Analysis to design and project engineers. In particular, design and quality engineers alike will appreciate the fact that the analysis techniques described herein are the result of several years of successful application of worst case analysis methodology to "real world" design engineering projects.

1.3
TRADITIONAL WORST CASE ANALYSIS

Worst Case Analysis is a thorough and rigorous mathematical process which determines the minimum and maximum values of each system parameter of concern, including each system function (gain, velocity, response time, etc.) and component stress (temperature, power, pressure, etc.).

WCA examines all of the component, environmental, and interface variables which affect system performance; e.g., initial tolerances, ambient temperature, altitude, inputs, loads, power sources, etc., and determines their combined effects. The min/max "worst case" values are compared against allowable specification limits to determine whether the system will meet its functional requirements, and whether any of the components in the system will be damaged or degraded:

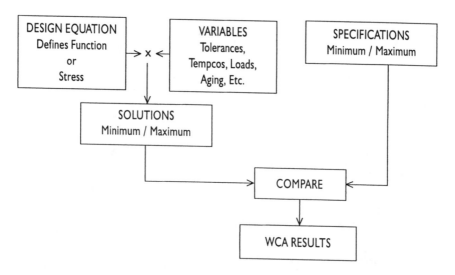

Figure 1.3-1
WCA Flow Diagram

A key benefit of WCA methodology is the discipline which it imposes upon the design validation process, which ensures that designs are fully and carefully evaluated.

1.4
ENHANCED WORST CASE ANALYSIS (WCA+)

A justified criticism of traditional WCA is that the worst case values identified by the analysis may occur so infrequently as to be negligible, resulting in costly overdesign. Several years ago Design/Analysis Consultants, Inc. (DACI) addressed this issue by developing an enhanced worst case analysis methodology called Worst Case Analysis *Plus* (WCA+^{TM}). In addition to generating traditional min/max solutions, WCA+ methodology includes the calculation of probabilities and sensitivities which are used for risk assessment and design optimization:

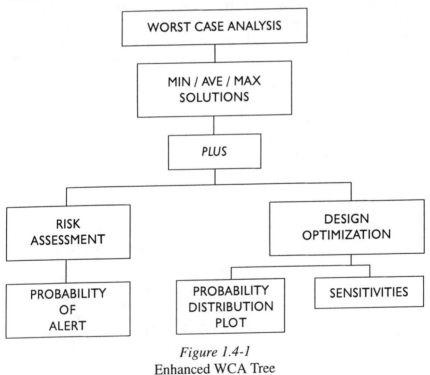

Figure 1.4-1
Enhanced WCA Tree

After identifying the min/max limits of performance, if an out-of-specification (ALERT) condition is detected, the data collected during the analysis can be used to generate a probability distribution plot:

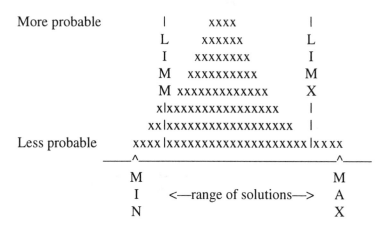

Figure 1.4-2
Probability Distribution Plot

In the example above, the x's mark the probability distribution. As the equation's solution varies from its minimum (MIN) to its maximum (MAX) value, the height of the distribution indicates the probability of the solution. The total area of a probability distribution equals one. Therefore, the total area outside the specification markers (LIMM = minimum spec limit; LIMX = maximum spec limit) equals the fraction of the time that the equation's solutions will not meet specification.

The probability distribution plot provides design and quality managers with an effective tool for rational risk assessment and design centering, helping to ensure reliable and cost-effective designs.

Design centering is facilitated by the identification of parameters which have the greatest effect on the design, commonly referred to as sensitivity analysis. Like probabilities, sensitivities can also be efficiently generated during the worst case analysis. Sensitivities allow the designer to optimize the design by adjusting the more significant variables in the direction which shifts the design center in the desired direction. In the distribution above, the center of the range of solutions is to the left of optimum. After design centering, the range of solutions will be centered between the specification limit "goal posts":

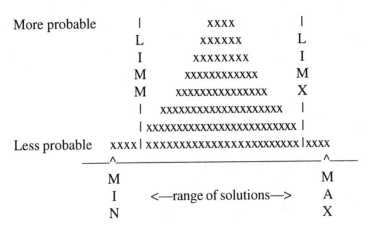

Figure 1.4-3
Probability Distribution Plot, Centered

In summary, enhanced Worst Case Analysis combines the traditional benefits of WCA with the added benefits of probability and sensitivity analyses. WCA+ is a powerful design validation methodology which yields optimized and highly reliable designs.

1.5
KEY DEFINITIONS

Definition # 1: Worst Case Analysis

Worst Case Analysis (WCA) is the mathematical determination of the limits of possible performance of a system to ensure that the desired system responses will not exceed specification limits, and that none of the system components will be damaged or degraded.

WCA consists of three sub-analyses, the Functional Margins Analysis, the Stress Margins Analysis, and the Applications Analysis:

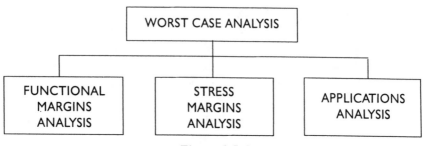

Figure 1.5-1
Key Elements of Worst Case Analysis

Definition # 2:   Functional Margins Analysis

A *Functional Margins Analysis* is the calculation of the minimum and maximum values for each specified system function. The min/max values are compared to the corresponding allowable limits as defined by the system specification. Values which exceed specification limits are identified as ALERT cases; those which are within specification limits indicate acceptable performance (OK category).

The Functional Margins Analysis (FMA) identifies those portions of the design where system performance exceeds specification limits; i.e., a "tolerance" analysis. Functional parameters and limits are usually defined by the equipment specification. Examples of functional parameters are gain, bandwidth, acceleration, flow rate, focal point, mechanical dimensions, etc.

Definition # 3:   Stress Margins Analysis

A *Stress Margins Analysis* is the calculation of the maximum applied stress for each stress parameter for each component in a system. The maximum stress value is compared to the corresponding allowable stress limit as defined by the component specification. Stress values which exceed stress limits are identified as ALERT cases; those which are within stress limits indicate acceptable stress (OK).

The Stress Margins Analysis (SMA) identifies those portions of the design where absolute maximum limits are exceeded; i.e., a "breakage" analysis. Stress parameters and limits are usually defined under the heading "Absolute Maximum Ratings" on component data sheets. Examples of stress parameters are voltage, pressure, power, shear, intensity, temperature, etc.

Definition # 4:   Applications Analysis

> An *Applications Analysis* is an evaluation of all applications information, characterization data, special test data, and other miscellaneous and supplementary data related to the components and subassemblies employed in a design. Application violations are identified as ALERT cases.

The Applications Analysis is a "catch-all" supplementary analysis which is used to identify important design constraints which are not documented in the primary specifications. For example, a footnote at the bottom of an applications note may contain this important restriction: "A load capacitance exceeding 50 picofarads may cause oscillations."

1.6
ORDER OF ANALYSIS

The order of analysis is (1) Functional Margins Analysis, (2) Stress Margins Analysis, and (3) Applications Analysis. The FMA is performed first since its results can be used as input data for the SMA. For example, the FMA of an electronics system will identify minimum and maximum voltages and currents; these values are then used in the SMA to determine maximum component stresses. The Applications Analysis is intended to identify subtle design defects, and is best performed after a thorough understanding of the design has been obtained via the FMA and SMA.

1.7
MISCONCEPTIONS

Misconception # 1:   A Worst Case Analysis Isn't Really Necessary If We Have A Good Test Plan

Testing alone is not sufficient for design validation. In general, it is simply not practical to test prototype units for the full range of conditions which may be experienced by a product when it leaves the factory. Also, the prototypes themselves will embody only a very small sampling of the enormous possible combinations of component tolerances. Therefore, worst case performance will almost certainly not be detected by prototype testing, which means that design deficiencies may not be identified.

The above observations are not meant to downplay the importance of testing; testing is *essential.* The point is that analysis complements a test program by identifying problems that even the most rigorous testing can miss:

DESIGN VALIDATION RULE # 1
TESTING + ANALYSIS = ENGINEERING EXCELLENCE

Misconception # 2: A Worst Case Analysis Is Too Pessimistic...
Its Results Will Never Occur

Worst case results can and do occur, as the case histories presented later demonstrate. Also, small probabilities of occurrence can be significant. For example, a five percent failure rate may consume all or more of a product's profits in warranty repair costs; a 0.1 percent failure rate may not be acceptable for a critical safety application.

Some analyses do give unrealistically pessimistic results, i.e., "worse than worst case," because the supporting calculations do not include the effects of dependent variables. By definition, a proper worst case analysis will account for the effects of dependent variables. And finally, a comprehensive analysis methodology will include an estimate of the probability of occurrence of identified problems, so the significance of the problems can be rationally evaluated.

Misconception # 3: A Computer Simulation Is A Worst Case Analysis

While simulation is an important and cost effective means of determining a design's basic functionality, simulation should not be confused with analysis. Simulations are actually a form of "computer prototyping," and in that sense can be considered an extension of the test process.

Simulation tends to discourage a full understanding of system operation by allowing easy tinkering with system parameters, equivalent to tweaking a prototype until a desired response is obtained. Such "tweak and tune" experimentation usually does not expose worst case performance limits or system sensitivities, and can leave in place hidden bugs which will later manifest themselves at great cost. Also, complex simulations employing large amounts of data entry invite errors by decoupling the designer from the system details, which leads to the complexity paradox:

THE SIMULATION COMPLEXITY PARADOX

If a simulation requires complex models to achieve believable results, believable results are harder to achieve because of the need for complex models.

Although simulation techniques are not in general the optimum choice for worst case analysis tasks, simulations are routinely used (along with test results) to validate the basic functionality of a design. This allows WCA efforts to focus on performance limits rather than function, which can reduce analysis costs.

For example, it is common practice for the logic functions of a digital electronics system to be validated by a logic simulation. WCA is then used to validate the setup and hold margins, which are key timing constraints that can be evaluated independently. In other cases, the evaluation of limits is irreducibly intermingled with functional performance, so WCA evaluates both limits and functionality simultaneously. In such cases, therefore, it is possible to omit the simulation.

Misconception # 4:   Our Engineers Perform Their Own WCAs

The WCA should be implemented by independent and objective analysts. This guards against the blind spots and prejudices that exist in even the most talented design team. Also, engineers are often "under the gun" to meet schedules. This pressure tends to cause engineers to analyze only the perceived areas of risk; areas which are assumed safe or low-risk are skipped or only given cursory attention. As case histories demonstrate, some seemingly trivial circuits contain bugs which result in enormous corrective action costs.

An independent analysis also offers additional insight into the design, which increases the proficiency of the design team. Also, if performed by a specialist, an independent analysis will generally be generated much more efficiently, resulting in cost savings.

Other professions routinely employ independent checks on performance, from proofreaders to copilots. The engineering profession should too.

Misconception # 5: Our Field Experience Proves That WCA
Is Not Necessary

Design deficiencies may not exhibit themselves for years. When components are purchased for assembly, they may come from a lot which has its parameter values skewed heavily toward one end or the other of the allowable range. Over time, if enough parts are purchased from the same vendor, average parameter values will tend to be at the center of the specification. Such averaging may not occur until several lots have been produced and shipped by the vendor; i.e., a true average may not be experienced until several years have passed.

Therefore, it is possible that large numbers of a product may be initially produced with components which are not representative of the full range of possible component values. It is quite possible (as has been observed during problem investigations) for "inexplicable" failures (i.e., design-related failures due to tolerance buildup) to start to occur in the third or fourth year of production, sometimes at an alarmingly high frequency. The consequence is:

THE PRODUCTION MANAGER'S RELIABILITY RULE
Favorable field data do not guarantee future success.

"Build-to-print" manufacturers are often the victim of this rule, and should protect themselves as follows: (1) always ask for the Worst Case Analysis which supports the design (there is often no analytical data available, which indicates high risk), (2) insert a clause in the contract that will protect against the consequences of hidden design defects, (3) if expensive problems arise which are not manufacturing-related, employ design experts to generate the proof needed to demonstrate a faulty design.

1.8
COST EFFECTIVENESS

Without effective design analysis, it is a rare program that does not see its development schedule stretched and budget expanded by the unsettling emergence of previously undetected design bugs. Likewise, production cycles often suffer several interruptions due to the need for design "improvements," which are really corrections of design defects which sneaked through the prototype evaluation process. If latent bugs escape into the field, they can cause warranty costs to skyrocket and the company's reputation to be damaged.

A review of data from a large sample of DACI's worst case analysis services (1990-1991) yielded impressive results: WCA generates savings which are more than *seven times its cost* on average.

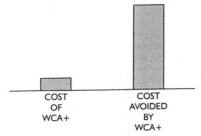

Figure 1.8-1
Worst Case Analysis Cost Effectiveness

The analyses sampled were for designs which had already reached the prototype stage with satisfactory results. Therefore each defect detected by WCA had not been previously detected by test or simulation activities. The data review assumed the cost of correcting an undiscovered design defect to be $5000/bug (1991), but this is believed to be very conservative. Project managers in the electronics industry have advised DACI that it costs on the order of tens of thousands of dollars to correct a single simple bug if it is detected during the production phase. The cost can be significantly higher if the bug is serious enough to shut down the production line. If the bug is detected during or following system installation or delivery, the related warranty costs can be very high; consequential costs related to damage to the company's reputation and loss of future business can be catastrophic.

The calculated savings did not include supplementary benefits which accrue from a better understanding of the design. Also, the savings reflected DACI's extensive WCA experience and efficiency which may result in lower analysis costs than will be obtained by less experienced analysts.

1.9
CASE HISTORIES

Case History # 1:   The Case of the Missing Methodology

A critical circuit required for a military system would not work over the required temperature range. Extensive computer simulations, testing, and design reviews failed to identify the problem, and out-of-spec performance continued to occur on the prototype units. A mathematical analysis finally identified the problem as a subtle design defect related to a parameter which was not accounted for in the simulations. When the problem was identified the solution became apparent, but it came too late to meet a critical schedule: the project was cancelled.

In Case 1, in spite of intensive simulation and test efforts, the omission of analysis as a design validation tool allowed a significant design parameter to be overlooked until it was too late, with devastating effects on a critical military program.

Case History # 2:   The Case of the Design-Change Disaster

Sometimes an adequate design is circumvented by design changes downstream which are not analyzed for their worst case effects. The collapse of the walkways in the Kansas City Hyatt Hotel in 1981 is a case in point.

As described in *Flying Buttresses, Entropy, and O-Rings* by James L. Adams (Harvard University Press, 1991), the Hyatt walkways collapsed during a dance, resulting in the loss of 114 lives. The walkways, which consisted of upper and lower sections, had been originally designed as beams suspended from a series of rods, with both the upper and lower walkways connected to each rod by threaded nuts/washers. This arrangement allowed each nut/washer on a rod to support one walkway load.

Prior to construction it was determined that assembly would be simpler if the rods were split into two separate sections, using the top section to first suspend the top walkway, and then using the bottom section to hang the lower walkway from the top walkway. The modified design was implemented without being analyzed, which would have revealed that the new arrangement resulted in twice the original load being placed on the upper rod's nut/washer.

During the dance, one upper nut/washer pulled through its beam. This failure placed additional load on adjacent rods, causing a chain reaction of failures which resulted in the collapse of the walkways.

A major benefit of WCA is the discipline that it imposes upon the design process. A company policy to routinely apply WCA to all design activities, including design changes, is a powerful antidote to the unintended introduction of costly errors.

Case History # 3:   The Case of the Insensitive Sensor

A sensor unit was successfully produced in small quantities, but when put into production many fielded units would drift out of specification. An analysis determined that there was insufficient design margin and that a large percentage of units produced would not be capable of meeting requirements.

In Case 3, a unit had been "successfully" developed and produced in small quantities. Following full-scale production and subsequent delivery of many units, an unacceptable failure rate developed. A review of the development records found no formal analysis report. Instead, analysis documentation consisted solely of inadequate informal calculations recorded in an engineering notebook.

Case History # 4:   The Case of the Fickle Feedback (An Example of the Simulation Complexity Paradox)

A custom analog integrated circuit was designed with a rather complex internal feedback network. Design validation was by a matrix of computer simulations which was intended to check "worst case" parameters; i.e., integrated circuit process variables, temperature, and supply voltage. The simulations showed proper performance.

The initial prototype ICs failed to function properly, and the failure was suspected to be related to the feedback network. Additional simulations "confirmed" the hypothesis. Based on simulation results, the design was modified, and simulations were again used to validate the redesign. The redesigned ICs also failed, with the failure again determined to be related to the feedback network. The design effort was terminated.

Case 4 is an example of the pitfalls of over-reliance on simulation. During the prototype problem investigation process, it was pointed out to the IC designer that, since the prototype performance of the ICs did not match simulation results, simulations should not continue to be used to predict future performance; i.e., the Simulation Complexity Paradox. It was recommended that no design corrections be finalized based on simulation data until the prototype simulation models were adjusted so that they would correlate with prototype performance. This advice was not followed.

Case History # 5:   The Case of Save a Dime, Spend a Dollar

The program manager pled with upper management to allow WCA to be performed on his system which was soon to be installed. He wanted to be sure the system worked well since it was going to cost $50,000/week to install the system, and the analysis would not have cost more than $25,000. Management said no. The program manager's fears were realized as several hidden design problems manifested themselves, which stretched the installation time out by three months, or about $650,000. Following installation intermittent problems continued to occur.

Case History # 6:   The Case of the $5 Million 1/4 Watt Resistor

An industrial electronics product was not subjected to WCA. After a high volume of product was shipped, an underrated 1/4 watt resistor used in the product failed, resulting in a five million dollar corrective action effort.

Case History # 7:   The Case of the Presumed Innocent

A circuit was designed using +/- 1 percent initial tolerance resistors, and "by inspection" was presumed to be adequate for the application. A Worst Case Analysis of the circuit determined that an ALERT would occur with a probability greater than 46 percent.

Case 7 demonstrates how the discipline and thoroughness imposed by WCA methodology will catch problems in areas dismissed by "intuitive" thinking.

Case History # 8:   The Case of the Prior Perfection (or, "It's Always Worked Before")

A producer of radios was experiencing a large number of circuit card failures. The company had previously produced the same card in large volume without problems. Low-gain transistors were suspected, and extensive parts testing was performed. A subsequent analysis determined that the design was deficient and the "low gain" parts were well within specification.

Case 8 demonstrates the folly of relying primarily upon field experience to validate a design. It is important to recognize that design deficiencies may not exhibit themselves for years due to favorable tolerances, and then suddenly exhibit themselves in overwhelming numbers in later years due to unfavorable tolerances. These deficiencies could have been detected with WCA. In Case 8 for example, the design was initially assumed to be correct since the program was "build-to-print" and hundreds of units had been previously fielded. An investigative analysis was performed which identified the existence of a design deficiency. The faulty notion that the design was OK resulted in a lot of wasted hours and unnecessary finger-pointing at an innocent parts vendor.

Case History # 9:   The Case of the Troubled Troubleshooting

A newly-developed product had a problem meeting specification during initial production. It was observed that by swapping out a part, the problem disappeared. The removed "bad" part was analyzed to determine what made it different from the newly-installed "good" part that made the unit work. Some parameter differences were identified, and the part spec was modified to ensure that only "good" parts were installed in the product.

After shipping a few dozen units, the problem returned, and trouble-shooting resumed. By swapping parts, a different part was found to have caused the problem. The newly-identified "bad" part was again compared to the newly-installed "good" part, its spec was modified, etc., etc., and so it went. A few hundred thousand dollars later, after the umpteenth part swap fix fizzled, it was decided to try something novel: understand the problem. Sure enough, a worst case analysis identified a weakness in the basic design that had nothing to do with the previously identified "bad" parts.

Moral: Testing *plus* analysis is the key to troubleshooting success.

CHAPTER TWO —————————————

HOW TO PERFORM A WORST CASE ANALYSIS

Enhanced Worst Case Analysis methodology consists of 10 key steps:

STEP 1:   Define the Analysis Requirements
STEP 2:   Establish Derating Factors
STEP 3:   Determine Global Parameters
STEP 4:   Develop the Design Equations
STEP 5:   Identify the Worst Case Solutions
STEP 6:   Determine the Margins
STEP 7:   Calculate ALERT Probabilities of Occurrence
STEP 8:   Reconciliation and Feedback
STEP 9:   Perform a Risk Assessment
STEP 10: Correct and Optimize the Design

2.1
DEFINE THE ANALYSIS REQUIREMENTS

The scope and depth of the analysis should be agreed upon in advance by the design team in conjunction with company experts and/or outside consultants. General analysis requirements should be defined by an applicable WCA standard (see Appendix B). Specific cases should be derived from the system specifications, or based upon customary engineering standards or practices when explicit specifications are not available.

Analysis by definition demands thinking, which requires that the analysis be broken into small comprehensible "chunks," or analysis cases. The cases are defined by an Analysis Table as in the example below:

ANALYSIS TABLE for XYZ PROJECT

CASE/DESC	UNITS	——SPEC——		———ACTUAL———			STATUS
		Minimum	Maximum	Minimum	Average	Maximum	
1.0 AMPLIFIER #1							
1.1 DC Gain	V/V	185	215				
1.2 Bandwidth	Hz	9000	11000				
2.0 ACTUATOR							
2.1 Response time	mS	0	1.5				

Etc.

The Analysis Table is generally used to define only the Functional Analysis requirements. The Stress Analysis by definition is required to analyze *all* stress parameters for *all* components. Likewise, the Applications Analysis addresses *all* applications and miscellaneous supporting data. Therefore an analysis plan for these two analyses is superfluous. However, the Analysis Table can be expanded and used to summarize the results of the entire analysis.

2.2

ESTABLISH DERATING FACTORS

Derating factors are employed to help account for uncertainties in the design. For example, when evaluating the stress on a part, a derating factor of ten percent may be applied to the data sheet limit. This means that the worst case maximum stress applied to the part should not exceed 90 percent of its actual limit; i.e., a ten percent safety margin has been added to the design. A sample Derating Guideline for electronics equipment is shown below:

PART TYPE	PART STYLE	PARAMETER	FACTOR
CAPACITOR	Aluminum	RMS amps	0.80
CAPACITOR	Aluminum	DC volts	0.95
DIODE	Rectifier	Forward amps	0.90
DIODE	Schottky	Reverse volts	0.85
IC	Linear	Input volts	0.90
RESISTOR	Film	Power, watts	0.90
TRANSISTOR	Bipolar	Junction temperature, C	0.95

Etc.

Derating factors will vary depending upon the application, with non-critical applications (e.g., office equipment) having little or no derating, while critical applications (e.g., aircraft) may employ significant derating. Derating factors may also be applied to help ensure longer operating lifetimes by reducing wear-out stresses, particularly for mechanical assemblies. Lifetime extension is not applicable for some part types, e.g., most electronic components, because inherent reliability is so good as to constitute "lifetime" operation. For such part types, deratings should only be used to compensate design uncertainties.

IMPORTANT: In the absence of uncertainties or the need for extended lifetimes, the use of derating factors will result in overdesign and associated higher costs. Therefore derating factors should be applied judiciously. The employment of WCA+ risk assessment methodology will significantly reduce uncertainties in the design, allowing the use of derating factors to be reduced or eliminated in many cases.

2.3
DETERMINE GLOBAL PARAMETERS

Identify the interface and environmental parameters which will be used throughout the analysis; e.g., temperature limits, power source limits, aging tolerances, etc. Defining global parameters at the start of the analysis ensures that separate sections of the analysis will employ a uniform baseline. A sample global library is provided below:

GLOBAL VARIABLES

	NAME	MIN	MAX	DESCRIPTION
1	Ta	0	75	Ambient temperature range, C
2	V5	4.75	5.25	+5V power supply, volts
3	V12	11.4	12.6	+12V power supply, volts
4	TOLag	-1.00E-3	1.00E-3	+-1000 ppm/year aging tolerance

Etc.

2.4
DEVELOP THE DESIGN EQUATIONS

By obtaining or developing and then examining relevant design equations, a designer can visualize how a system's performance is affected by variations in parameters. This provides important insight into system operation, and allows performance to be optimized. Also, equations are amenable to the application of statistical methods which allow the probabilities of occurrence of detected faults to be estimated.

Design equations are generally obtained from the following sources:
- textbooks
- math-based software
- applications literature
- original work previously documented in engineering notebooks and reports

For complex or unfamiliar equations, it is recommended that at least two independent data sources be used to validate their accuracy.

Obtaining relevant design equations can require a moderate amount of research. Even more time consuming, however, is the effort required to collect the data which define the equation's variables; e.g., initial tolerances, temperature coefficients, transfer functions, absolute maximum ratings, aging factors, etc. Some data, such as aging factors, are very difficult to obtain and may have to be estimated. Such estimates and all other parameter assumptions should be clearly identified in the analysis.

2.5
IDENTIFY THE WORST CASE SOLUTIONS

The identification of worst case solutions requires that the analyst determine the values of the equation's variables which will cause the equation to have a minimum and maximum value. The basic methods for accomplishing this are described below. First, however, the concept of *monotonicity* is defined. As will be seen, a knowledge of the monotonicity of an equation with respect to each of its variables can greatly reduce the number of calculations required to identify worst case solutions.

Definition # 5:   Monotonicity

> An equation is monotonic with respect to a given variable V if there are no peaks or valleys in the equation's solutions as the variable is varied from its minimum (Vm) to its maximum (Vx) value.

Mathematically, if the derivative (slope) of an equation with respect to a given variable is not equal to zero over the variable's range of values, then the equation is monotonic with respect to the variable. It follows that the minimum and maximum values of an equation with respect to a given monotonically-related variable will occur at the end-point values of the variable. Some examples of monotonic equations are shown below:

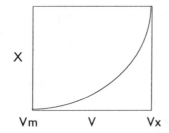 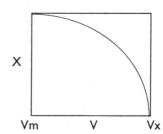

Figure 2.5-1
Examples of Monotonic Equations

The significance of monoticity is that only the min/max values of the variable (Vm, Vx above) need to be used for worst case calculations; i.e., the required number of calculations can be minimized:

Definition # 6: Evaluating Monotonic Equations

Equations with monotonically-related variables should be evaluated at the min/max values of all variables.

If the equation is non-monotonic, then it follows that a peak or valley value may exist for the equation within the range of V:

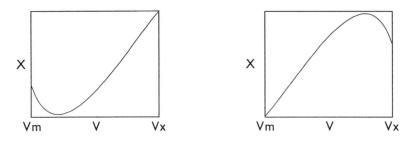

Figure 2.5-2
Examples of Non-Monotonic Equations

However, even if X contains peaks or valleys with regard to V, this does not mean that the min/max solutions of X will occur at the peaks or valleys, because inflections may occur:

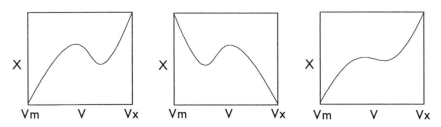

Figure 2.5-3
More Non-Monotonic Examples

The conclusion is:

Definition # 7:   Evaluating Non-Monotonic Equations

> Equations with non-monotonically-related variables should be evaluated at all zero-slope points in addition to the min/max values of the variables.

2.5.1   Identifying Worst Case Solutions by Inspection

EXAMPLE: $X = A/B$; $A = 1$ to 3; $B = 2$ to 4

By inspection, the minimum value of X will occur for the minimum value of A and the maximum value of B; the maximum value of X will occur for the maximum value of A and the minimum value of B:

Xmin = 1/4   (Amin, Bmax)

Xmax = 3/2   (Amax, Bmin)

2.5.2   Identifying Worst Case Solutions by Differentiation

EXAMPLE: $X = A/B$; $A = 1$ to 3; $B = 2$ to 4

Using the prior simple example to illustrate, the derivative of X with respect to A is 1/B. However, 1/B is not equal to zero for any value of B. Therefore, X is monotonic over the range of A, which means that the minimum or maximum value of A will correspond to the minimum or maximum value of X. Likewise, X is monotonic with respect to variable B ($dX/dB = -A/B^2$; not equal to zero for any value of A or B). At this point the analyst knows that the minimum and maximum values of A and B will yield the minimum and maximum values of X, but does not know which combinations of A and B are required; i.e., four calculations are required:

Amin /Bmin = 1/2
Amax /Bmin = 3/2
Amin/Bmax = 1/4

Amax/Bmax = 3/4

From the above, the worst case results are

Xmin = 1/4   (Amin, Bmax)
Xmax = 3/2   (Amax, Bmin)

This result is of course the same as determined by inspection in the prior example.

EXAMPLE: X = 3*SIN(wt), wt = 0.67 to 2.56, radians

The derivative of X with respect to wt is 3*COS(wt). Setting the derivative to zero and solving for wt:
3*COS (wt) = 0
wt = Pi/2 = 1.5708 radians

Therefore a minimum or maximum value of X will occur at wt = Pi/2, which is within the range of values of wt. The possible solutions of X are therefore:

3*SIN(0.67) = 1.863   (using min value of variable wt)
3*SIN(2.56) = 1.648   (using max value of variable wt)
3*SIN(Pi/2) = 3.0   (using zero-slope value of variable wt)

From the above, the worst case results are

Xmin = 1.648 (wt = 2.56)
Xmax = 3.0   (wt = Pi/2)

2.5.3 Identifying Worst Case Solutions by Brute Force

Calculate all equation solutions corresponding to the range of values of all variables and identify the minimum and maximum solutions. If it is known that the equation is monotonic with respect to a given variable, then it is only necessary to use the minimum and maximum values of the variable when calculating solutions. If the equation is not monotonic with respect to a given variable, then there must be enough sample points for the variable to ensure that peaks and valleys in the equation's solutions will be identified.

EXAMPLE: The minimum and maximum peak height of a projectile:

The following expression defines a projectile's height H versus time, neglecting air resistance:

$$H = Vo*SIN(0.7854)*t-0.5*32.17*t^2$$

where Vo = starting velocity, ft/sec = 250 to 750 ft/sec
0.7854 radians = starting angle (45 degrees)
32.17 = gravitational acceleration, ft/sec^2
t = time, seconds = 0 to 35 seconds

The graph of H versus t is shown below:

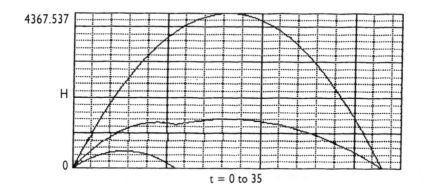

Figure 2.5.3-1
Projectile Height Graph

Since H is non-monotonic with respect to t, the minimum and maximum peak height must be determined indirectly from an inspection of the graph; i.e., the values are dependent upon the resolution of the calculations, which is proportional to the number of sampling points used for t. In this example, 36 sampling points were used for t, and 2 sampling points were used for Vo, for a total of 72 calculations. From the graph above, the min/max peak height is estimated to be 500 to 4368 feet.

It is preferable to employ a monotonic expression which directly expresses the projectile peak height; i.e.,

Hpk = (Vo*SIN(.7854))^2/(2*32.17)

Evaluating the above expression requires only two calculations corresponding to the min and max values of Vo. The min/max peak height is directly determined to be 485.7 to 4371.3 feet.

In the above example, the monotonic equation allowed the precise worst case values to be obtained in two calculations; the non-monotonic equation required 72 calculations and yielded only approximate solutions.

EXAMPLE: Noninverting voltage feedback amplifier gain:

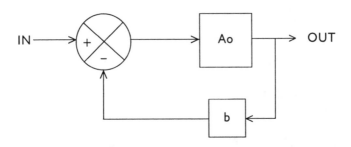

Figure 2.5.3-2
Noninverting Feedback Amp

In this example, it is desired to determine the gain of the amplifier. From the Analysis Table (see Section 2.1), a gain of 185 to 215 is required (200 +/-7.5%):

CASE/DESC	UNITS	———SPEC———		———————ACTUAL————————			STATUS
		Minimum	Maximum	Minimum	Average	Maximum	
1.1 DC Gain	V/V	185	215				

Analysis Case 1.1 requires a closed loop gain of 185 minimum to 215 maximum volts/volt. Calculations are to be recorded on a worksheet. The worksheet may be a form, a spreadsheet, or generated with the assistance of WCA software; the example below was generated using DACI's *Design Master*™ software. DC gain is identified by the symbol "Adc" which is the left-hand side of the equation, or the unknown. The right-hand part of the equation (the math expression) is defined by the standard equation for a noninverting amplifier:

	SYMBOL	DESCRIPTION	EXPRESSION
1	Adc	DC Gain	Ao/(1+Ao*b)

The variables Ao (open loop gain) and b (feedback ratio) are themselves functions of variables (i.e., they are dependent variables). They are defined in supporting equations:

	SYMBOL	DESCRIPTION	EXPRESSION
2	Ao	Open loop gain	10000*(1+TcA*(Ta-25))

	SYMBOL	DESCRIPTION	EXPRESSION
3	b	Feedback ratio	R1/(R1+R2)

Dependent variables R1 and R2 have been used to define b, so they also must be defined with supporting equations:

	SYMBOL	DESCRIPTION	EXPRESSION
4	R1	Inverting input resistor	1E3*TiR1*(1+TcR1*(Ta-25))

	SYMBOL	DESCRIPTION	EXPRESSION
5	R2	Feedback resistor	199E3*TiR2*(1+TcR2*(Ta-25))

Independent variables TcA, Ta, TiR1, TcR1, TiR2, and TcR2 are entered on a separate portion of the worksheet. Note that even though R2's tolerances have the same value as R1's tolerances, they must be separately

– 34 –

defined since R1 and R2 are independent components:

INDEPENDENT VARIABLES:

NAME	MIN	MAX	POINTS	DESCRIPTION
1 TcA	-1.00E-4	1.00E-4	2	Tempco of amp, +-100 ppm/C
2 Ta	0.0	75	2	Ambient temperature range, C
3 TiR1	0.98	1.02	2	Initial tolerance of R1, +- 2%
4 TcR1	-3.00E-4	3.00E-4	2	Tempco of R1, +-300 ppm/C
5 TiR2	0.98	1.02	2	Initial tolerance of R2, +- 2%
6 TcR2	-3.00E-4	3.00E-4	2	Tempco of R2, +-300 ppm/C

MIN and MAX are the minimum and maximum values of the variable. The POINTS entry defines the number of sampling points over the range. In this example, since the analyst has experience with the standard equation and knows it to be monotonic with regard to its variables, POINTS is set to 2; i.e., only the minimum and maximum values of each independent variable are used during computations.

If a variable were non-monotonic with respect to the equation, or if its monotonicity were unknown, the sampling points for the variable would need to be increased sufficiently to resolve any peaks or valleys in the range of solutions.

After setting up the worksheet as described above, a calculation table is created, with a calculation required for each combination of variables; i.e.,

$$N = (V1 \text{ samples}) \times (V2 \text{ samples}) \times ... (Vn \text{ samples})$$

where N = total number of required calculations
Vn = number of sample points for nth variable

In the example above, there are 6 independent variables with 2 sample points each, therefore

$$N = 2 \times 2 \times 2 \times 2 \times 2 \times 2 = 2^6 = 64 \text{ calculations}$$

For each combination of variables, the solution to the primary expression is calculated. Note that it is imperative that the dependent variables as defined by the supporting equations be absorbed into the primary equation, otherwise unrealistic results (i.e., worse than worst case) will be obtained.

– 35 –

After absorbing the supporting equations, the primary equation is stated in terms of constants and independent variables only; i.e.,

$$Adc = Ao/(1+Ao*b)$$

$$= \frac{10000*(1+TcA*(Ta-25))}{1+(10000*(1+TcA*(Ta-25)))*(1E3*TiR1*(1+TcR1*(Ta-25)))/...}$$
$$...((1E3*TiR1*(1+TcR1*(Ta-5)))+(199E3*TiR2*(1+TcR2*(Ta-25))))$$

The above complex expression can of course be handled easily when constructed with spreadsheet or math software. A spreadsheet table of calculations is shown below:

CALCULATION TABLE for Adc

N	TcA	Ta	TiR1	TcR1	TiR2	TcR2	SOLUTION
1	-1e-4	0	0.98	-3e-4	0.98	-3e-4	196.088
2	1e-4	0	0.98	-3e-4	0.98	-3e-4	196.069
3	-1e-4	75	0.98	-3e-4	0.98	-3e-4	196.059
4	1e-4	75	0.98	-3e-4	0.98	-3e-4	196.098
5	-1e-4	0	1.02	-3e-4	0.98	-3e-4	188.581
6	1e-4	0	1.02	-3e-4	0.98	-3e-4	188.563
7	-1e-4	75	1.02	-3e-4	0.98	-3e-4	188.554
8	1e-4	75	1.02	-3e-4	0.98	-3e-4	188.589
9	-1e-4	0	0.98	3e-4	0.98	-3e-4	198.978
10	1e-4	0	0.98	3e-4	0.98	-3e-4	198.958
11	-1e-4	75	0.98	3e-4	0.98	-3e-4	190.404
12	1e-4	75	0.98	3e-4	0.98	-3e-4	190.440
13	-1e-4	0	1.02	3e-4	0.98	-3e-4	191.362
14	1e-4	0	1.02	3e-4	0.98	-3e-4	191.343
15	-1e-4	75	1.02	3e-4	0.98	-3e-4	183.112 (Min)
16	1e-4	75	1.02	3e-4	0.98	-3e-4	183.145
17	-1e-4	0	0.98	-3e-4	1.02	-3e-4	203.890
18	1e-4	0	0.98	-3e-4	1.02	-3e-4	203.869
19	-1e-4	75	0.98	-3e-4	1.02	-3e-4	203.858
20	1e-4	75	0.98	-3e-4	1.02	-3e-4	203.900
21	-1e-4	0	1.02	-3e-4	1.02	-3e-4	196.088
22	1e-4	0	1.02	-3e-4	1.02	-3e-4	196.069

23	-1e-4	75	1.02	-3e-4	1.02	-3e-4	196.059
24	1e-4	75	1.02	-3e-4	1.02	-3e-4	196.098
25	-1e-4	0	0.98	3e-4	1.02	-3e-4	206.893
26	1e-4	0	0.98	3e-4	1.02	-3e-4	206.872
27	-1e-4	75	0.98	3e-4	1.02	-3e-4	197.981
28	1e-4	75	0.98	3e-4	1.02	-3e-4	198.020
29	-1e-4	0	1.02	3e-4	1.02	-3e-4	198.978
30	1e-4	0	1.02	3e-4	1.02	-3e-4	198.958
31	-1e-4	75	1.02	3e-4	1.02	-3e-4	190.404
32	1e-4	75	1.02	3e-4	1.02	-3e-4	190.440
33	-1e-4	0	0.98	-3e-4	0.98	3e-4	193.239
34	1e-4	0	0.98	-3e-4	0.98	3e-4	193.221
35	-1e-4	75	0.98	-3e-4	0.98	3e-4	201.880
36	1e-4	75	0.98	-3e-4	0.98	3e-4	201.921
37	-1e-4	0	1.02	-3e-4	0.98	3e-4	185.839
38	1e-4	0	1.02	-3e-4	0.98	3e-4	185.822
39	-1e-4	75	1.02	-3e-4	0.98	3e-4	194.155
40	1e-4	75	1.02	-3e-4	0.98	3e-4	194.193
41	-1e-4	0	0.98	3e-4	0.98	3e-4	196.088
42	1e-4	0	0.98	3e-4	0.98	3e-4	196.069
43	-1e-4	75	0.98	3e-4	0.98	3e-4	196.059
44	1e-4	75	0.98	3e-4	0.98	3e-4	196.098
45	-1e-4	0	1.02	3e-4	0.98	3e-4	188.581
46	1e-4	0	1.02	3e-4	0.98	3e-4	188.563
47	-1e-4	75	1.02	3e-4	0.98	3e-4	188.554
48	1e-4	75	1.02	3e-4	0.98	3e-4	188.589
49	-1e-4	0	0.98	-3e-4	1.02	3e-4	200.929
50	1e-4	0	0.98	-3e-4	1.02	3e-4	200.909
51	-1e-4	75	0.98	-3e-4	1.02	3e-4	209.907
52	1e-4	75	0.98	-3e-4	1.02	3e-4	209.951 (Max)
53	-1e-4	0	1.02	-3e-4	1.02	3e-4	193.239
54	1e-4	0	1.02	-3e-4	1.02	3e-4	193.221
55	-1e-4	75	1.02	-3e-4	1.02	3e-4	201.880
56	1e-4	75	1.02	-3e-4	1.02	3e-4	201.921
57	-1e-4	0	0.98	3e-4	1.02	3e-4	203.890
58	1e-4	0	0.98	3e-4	1.02	3e-4	203.869
59	-1e-4	75	0.98	3e-4	1.02	3e-4	203.858

60	1e-4	75	0.98	3e-4	1.02	3e-4	203.900
61	-1e-4	0	1.02	3e-4	1.02	3e-4	196.088
62	1e-4	0	1.02	3e-4	1.02	3e-4	196.069
63	-1e-4	75	1.02	3e-4	1.02	3e-4	196.059
64	1e-4	75	1.02	3e-4	1.02	3e-4	196.098

From the array of possible solutions, the minimum and maximum values are 183.11 and 209.95 respectively. The average value is obtained by summing all the solutions and dividing by the number of solutions:

Average solution = sum of all solutions / number of solutions

$$= 12555.38/64 = 196.18$$

These results are entered into the Analysis Table:

CASE/DESC	UNITS	——SPEC——		————ACTUAL————			STATUS
		Minimum	Maximum	Minimum	Average	Maximum	
1.1 DC Gain	V/V	185	215	183.11	196.18	209.95	

2.6
DETERMINE THE MARGINS

For both the functional and stress analyses, the analyst first defines the required operating limits:

REQT = LOWER LIMIT <—range of acceptable values—> UPPER LIMIT

Worst case performance limits must fall within this range of values. Examples of acceptable ranges of performance include:

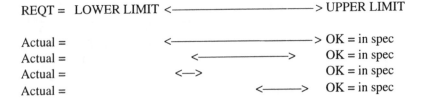

whereas examples of unacceptable ranges of performance include:

REQT =   LOWER LIMIT <————————————————> UPPER LIMIT

Actual =   <————————————————————————> ALERT
Actual =   <————————————————>   ALERT
Actual =   <—>   ALERT
Actual =   <—> ALERT

In a worst case analysis, the desired result is the operating margin, or the difference between the required limit and actual performance:

Definition # 8:   WCA MARGINS

Upper Margin = Max Limit (Required) - Max Limit (Actual)
Lower Margin = Min Limit (Actual) - Min Limit (Required)

For example, assume that a blower system is required to maintain an air flow of 100 +/- 5% cubic feet per minute (cfm), or 95 cfm (minimum allowable limit LIMM) to 105 cfm (maximum allowable limit LIMX). Assume that worst case calculations determine that the actual air flow can be 99 cfm (MIN) to 110 cfm (MAX):

```
/////////////////////L                    L ///////////////////////////////////
/////////////////////  I                  I  //////////////////////////////////
/////ALERT/////M                          M/////ALERT///////////////////
//////ZONE//////M                         X //////ZONE///////////////////////
////////////////////// |                  | ///////////////////////////////////
////////////////////// |                  | ///////////////////////////////////
```

	95	99	105	110	cfm

```
////////////////////// |           |        | /////////////// | ///////////
//////////////////—>| lower margin |<— —>|   upper margin   |<——//////
////////////////////// |           |        | /////////////// \///////////
////////////////////// |           M        | /////////////// M//////////
////////////////////// |           I        | /////////////// A //////////
////////////////////// |           N        | /////////////// X //////////
```

Figure 2.6-1
WCA Margins

In this example, the upper margin is 105 (spec) - 110 (actual) = -5. The negative result indicates an unacceptable (ALERT) condition. The lower margin is 99 (actual) - 95 (spec) = 4, which is a positive and acceptable (OK) condition. Since the upper limit is exceeded, however, the overall result is ALERT.

Assume that the design is changed such that the worst case actual air flow is now 99 to 104.5 cfm. The lower margin remains the same, and the upper margin becomes 105 (spec) - 104.5 (actual) = 0.5. Since both the upper and lower margins are positive, the overall margin is OK.

In the noninverting gain example of section 2.5.3, the upper margin is 215 (spec) - 209.95 (actual) = 5.05; status is OK. The lower margin is 183.11 (actual) - 185 (spec) = -1.89, which is an ALERT. The ALERT status is added to the Analysis Table:

CASE/DESC	UNITS	———SPEC———		————ACTUAL————			STATUS
		Minimum	Maximum	Minimum	Average	Maximum	
1.1 DC Gain	V/V	185	215	183.11	196.18	209.95	ALERT

2.7
CALCULATE ALERT PROBABILITIES OF OCCURRENCE

When a worst case result exceeds a specification limit, this does not necessarily mean that the design should be modified or scrapped. Indeed, such heavy-handed assessments will result in costly overdesign. Instead, a probability analysis should be performed on the analysis results to estimate the percent of the time that the specification limits will be exceeded. This approach allows design risk to be assigned in a rational way; e.g., it may be quite acceptable to ignore a design "flaw" which will only occur for 0.05% of units produced.

The term "statistical analysis" is used here in its general sense of inferring probability information from a sampling of data points. As will be shown, probabilities can be estimated without having to immerse oneself too far into the arcane trappings of statistical methodology. In fact, since the data required for a statistical analysis are generally not adequate for claims of high accuracy (hence the familiar lament, "lies, damned lies, and statistics"), the goals of this section are modest. Nonetheless, the designer or design manager should find the techniques presented herein immensely useful.

The primary goal of the probability analysis is to create a plot of the design equation's *probability distribution* (also referred to as *probability density*) as shown in Figure 2.7-1:

As the equation's solution varies from its minimum (MIN) to its maximum (MAX) value, the height of the distribution indicates the probability of the solution. Since the total area of a probability distribution equals one, the total area outside the specification markers (LIMM = minimum spec limit; LIMX = maximum spec limit) equals the fraction of the time that the equation's solutions will not meet specification. This fraction is referred to herein as the probability of an ALERT, or P(ALERT).

The probability distribution of a reliable, well-centered design equation will have none or only a small section of the distribution's "tails" sticking out past the specification goal posts, as in the figure above. An unreliable

The Design Analysis Handbook

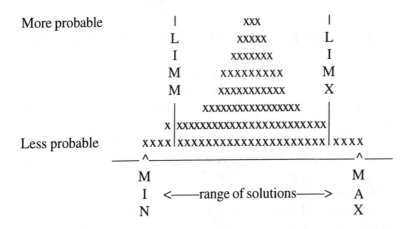

Figure 2.7-1
Design Equation's Probability Distribution

design will have a significant portion of the distribution extending beyond
the specification limits, as in the following example:

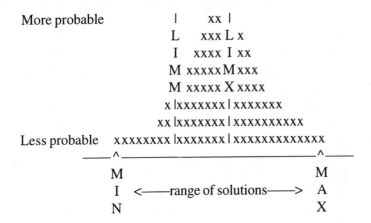

Figure 2.7-2
Probability Distribution of an Unreliable Design

The probability distribution is generated by weighting each solution in accordance with the combined probability distributions of the independent variables used in the equation. Techniques for accomplishing this will be described in the following sections, which will allow you to calculate P(ALERT) for your design equations.

2.7.1 Assigning Probability Distributions to Independent Variables

Each design equation has one or more independent variables, such as initial tolerances, temperature, temperature coefficients, supply voltages, loads, etc. Since many of these variables will only exhibit their extreme values a small percent of the time, a worst case result due to a combination of such variables will often have a very low likelihood of occurrence, which may reduce the need or extent of design modifications. Therefore it is very desirable to take advantage of variable distribution data.

Unfortunately, distributions are generally not known or are hard to obtain. Fortunately, the use of some conservative assumptions allows this obstacle to be readily overcome, as will be explained below.

Uniform Distribution

The uniform distribution describes a variable where any value within the variable's range is equally likely to occur.

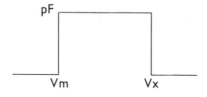

Figure 2.7.1-1
Uniform Distribution

where pF = probability function = height of distribution as a function of V
Vm = minimum value of V
Vx = maximum value of V

For the uniform distribution,
pF = 1/(Vx-Vm)

Note that the area under the curve is always equal to 1.0, which can be used to solve for the curve height pF.

If the range of variable V is sliced into S equal samples, the width of each sample is

$$pW = (Vx-Vm)/S$$

and the probability of the variable (pV) having the value of the sample point is

$$pV = pF*pW$$
$$= (1/(Vx-Vm))*(Vx-Vm)/S$$
$$= 1/S$$

Since all sample values of the uniform distribution are equiprobable, the probability of occurrence of a sample does not depend upon the min/max values of the variable or upon which slice contains the sample, but only upon the number of samples.

Because the extreme (min/max) values of the variable are as likely to occur as those in the center of the variable's range, the uniform distribution can often be used as a worst case approximation when the variable's actual distribution is not known.

The Triangular (Simpson) Distribution

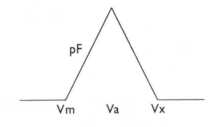

Figure 2.7.1-2
Triangular Distribution

Vm = minimum value of V
Va = average value of V
Vx = maximum value of V
pF = 4*(V-Vm)/(Vx-Vm)^2   for Vm <= V <= .5*(Vm+Vx)
   = 4*(Vx-V)/(Vx-Vm)^2   for .5*(Vm+Vx) <= V <= Vx

When S samples are taken, each sample is converted into a rectangular slice, with the sample point at the center of each slice.

For triangular distributions with S samples, the following approximation can be used to estimate the probability pV:

pV = 2*(V+dV-Vm)/((S+1)*(Va-Vm))   for V+dV <= Va

   = 2*(Vx-V+dV)/((S+1)*(Vx-Va))   for Va <= V+dV

where dV = .5*(Vx-Vm)/S

The sum of the probabilities across the range of S samples should equal 1.0. If the sum has a small rounding error, divide each pV value by the sum to correct the error.

For the triangular distribution, variable values near the center are much more likely to occur than those toward the min/max limits; this approximates the truncated normal distribution representative of many design variables. The triangular distribution can often be used as a simple "best case" approximation when the variable's actual distribution is not known.

The U Distribution (Bimodal)

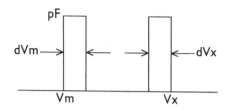

Figure 2.7.1-3
U Distribution

$$Vm = \text{minimum value of } V$$
$$Vx = \text{maximum value of } V$$
$$dVm = \text{width of left leg of U}$$
$$dVx = \text{width of right leg of U}$$
$$pF = 1/(dVm+dVx)$$
$$pV = (Vx-Vm)/(S*(dVm+dVx)) \text{ for values within dVm and dVx}$$
$$pV = 0 \text{ elsewhere}$$

An example of a variable with a U distribution is a low-precision component initial tolerance. Since it is not uncommon for manufacturers to sort components based upon initial tolerance, high-precision parts (closest to center value) will be selected from the group, leaving the remainder (the outer legs of the U) for the low-tolerance bin.

If variables with U distributions are present, the assumption of a uniform distribution being the worst case distribution will not hold, since the probability of occurrence of min/max values is higher for U-variables than for uniform variables. Therefore, the probability analysis should include U distribution formulas where applicable. An alternate approach for U-variables (or any complex distribution shape) is to break the variable's range into segments of approximately constant value, where the variable can be treated as a constant within the segment. An independent analysis is performed for each segment, with any resultant probabilities multiplied by the ratio of the segment's width to the total width of the distribution. After an analysis is performed for all segments, the factored probabilities are summed to get the total probability.

The Normal (Gaussian) Distribution

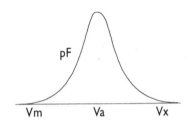

Figure 2.7.1-4
Normal Distribution

For the normal distribution,

$$pF = 1/(sdv*(2*Pi)^{\wedge}.5)*e^{\wedge}-((V-Va)^{\wedge}2/(2*sdv^{\wedge}2))$$

where sdv = standard deviation, constant > 0
        Va = average value of V

Since the normal distribution extends from minus infinity to plus infinity, to obtain a practical sample width (pW), the limits of the distribution must be truncated as shown below, where Vm = lower truncated value and Vx = upper truncated value. As noted earlier, it may be possible to obtain adequate results by using the simpler triangular distribution as an approximation to the normal distribution.

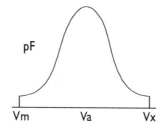

Figure 2.7.1-5
Truncated Normal Distribution

As noted above, the assumption of uniform probabilities generally results in a conservative P(ALERT) estimate. Therefore a uniform distribution can often be considered a "worst case" distribution. Since uniform distributions are coincidentally much easier to work with than non-uniform distributions, this leads to a useful guideline for a quick-and-easy preliminary probability estimate:

THE ANALYST'S RULE FOR A
QUICK-AND-EASY PROBABILITY ESTIMATE
Assume that all variables have uniform distributions

2.7.2 Using WCA Results for ALERT Probability Calculations

Explicit probability solutions are difficult to obtain, requiring the solving of multiple nested integrals. Even when using computer-generated numerical approaches, the required computation time can be prohibitive. Therefore, estimating procedures or special algorithmic software such as the *Design Master*™ are generally used for such calculations, which can yield reasonably conservative results fairly quickly. An example of a simple estimating procedure is described below.

The probability of occurrence of a specified result R out of N independent, mutually exclusive, and equiprobable results is

P = (Number of occurrences of R) / N

Since WCA is concerned with the probability of failure, this can be stated as

P(ALERT) = (Number of Out-of-Spec Results) / (Number of Results)

Since most design equations use variables with a continuous range of values, the expression above is only an estimate, with "number of results" being the number of independent sample solutions obtained for the expression. As the number of solutions increases the probability estimate becomes more accurate.

Using worst case methodology, solutions to an expression are calculated in an organized manner with each independent variable varied from its minimum to its maximum value while all other variables are held at a constant value. For each solution, the probability of occurrence of the solution is tabulated. After all solutions have been obtained, the probabilities of all ALERT solutions are summed to obtain the total probability of an ALERT for the expression.

Consider a function X which contains independent variables V1 through Vn. Each variable is incremented over its range a specified number of times (samples) to generate the total N number of solutions:

N = (V1 samples) x (V2 samples) x ... (Vn samples)

Assume that X = V1*V2, V1 = 2 to 4 using three samples, V2 = 0 to 10 using 3 samples, V1 and V2 have uniform distributions, and the specified allowable limits for X are 0 to 25:

VARIABLE	MIN	MAX	SAMPLE VALUES	# of SAMPLES
V1	2	4	2, 3, 4	3
V2	0	10	0, 5, 10	3

N = 3*3 = 9 total solutions

While performing the worst case calculations, out-of-spec results are tabulated to obtain the total number N(alert) of out-of-spec solutions:

LIMM=0 LIMX=25 Solution=V1*V2, V1 and V2=Uniform Distributions

N	V1	V2	SOLUTION	ALERT?
1	2	0	0	NO
2	3	0	0	NO
3	4	0	0	NO
4	2	5	10	NO
5	3	5	15	NO
6	4	5	20	NO
7	2	10	20	NO
8	3	10	30	YES
9	4	10	40	YES

From the above,

P(ALERT) = N(alert)/N = 2/9 = 0.2222 = 22.2 percent

In the above example, because of the small number of samples, the probability analysis provides a pessimistic result. To increase the accuracy of the probability estimate, the number of parameter value samples can be increased:

VARIABLE	MIN	MAX	SAMPLE VALUES	# of SAMPLES
V1	2	4	2, 2.5, 3, 3.5, 4	5
V2	0	10	0, 2, 4, 6, 8, 10	6

N = 5*6 = 30 total solutions

LIMM = 0   LIMX = 25   Solution = V1*V2, V1 and V2 = Uniform Distributions

N	V1	V2	SOLUTION	ALERT?
1	2	0	0	NO
2	2.5	0	0	NO
3	3	0	0	NO
4	3.5	0	0	NO
5	4	0	0	NO
6	2	2	4	NO
7	2.5	2	5	NO
8	3	2	6	NO
9	3.5	2	7	NO
10	4	2	8	NO
11	2	4	8	NO
12	2.5	4	10	NO
13	3	4	12	NO
14	3.5	4	14	NO
15	4	4	16	NO
16	2	6	12	NO
17	2.5	6	15	NO
18	3	6	18	NO
19	3.5	6	21	NO
20	4	6	24	NO
21	2	8	16	NO
22	2.5	8	20	NO
23	3	8	24	NO
24	3.5	8	28	YES
25	4	8	32	YES
26	2	10	20	NO
27	2.5	10	25	NO
28	3	10	30	YES
29	3.5	10	35	YES
30	4	10	40	YES

P(ALERT) = N(alert)/N = 5/30 = 0.1667 = 16.7 percent

As the number of samples is increased, the better the probability estimate P(ALERT) becomes. Although the number of samples would have to be infinite to obtain an exact probability, the use of worst case calculations guarantees that out-of-spec conditions are identified. Therefore, the worst case procedure described above ensures that the value obtained for P(ALERT) will be equal to or greater than the actual probability.

2.7.3   Using Non-Uniform Probability Distributions

When non-uniform distributions are used, it is necessary to add to the probability worksheet an additional column for each variable's probability (pVi) and a total probability column (pT). Since the variables are independent, their total probability is their product; i.e.,

pT = pV1 * pV2 * ... * pVn

Since all solutions are tabulated, the sum of all pT entries is equal to 1.0. To obtain P(ALERT), take the sum of the entries in the pT column which correspond to ALERT solutions; i.e.,

P(ALERT) = Sum of pT for all ALERT solutions

The probability columns are not used for uniform distributions only because they are trivial; i.e., uniform distributions allow a calculation short-cut. To illustrate, the prior example is repeated below using the complete probability calculations. For uniform distributions, pV1 = 1/5, pV2 = 1/6.

LIMM=0   LIMX=25   Solution=V1*V2, V1 and V2=Uniform Distributions

N	V1	V2	SOLUTION	ALERT?	pV1	pV2	pT
1	2	0	0	NO	0.200	0.167	0.03333
2	2.5	0	0	NO	0.200	0.167	0.03333
3	3	0	0	NO	0.200	0.167	0.03333
4	3.5	0	0	NO	0.200	0.167	0.03333
5	4	0	0	NO	0.200	0.167	0.03333
6	2	2	4	NO	0.200	0.167	0.03333
7	2.5	2	5	NO	0.200	0.167	0.03333
8	3	2	6	NO	0.200	0.167	0.03333
9	3.5	2	7	NO	0.200	0.167	0.03333

10	4	2	8	NO	0.200	0.167	0.03333
11	2	4	8	NO	0.200	0.167	0.03333
12	2.5	4	10	NO	0.200	0.167	0.03333
13	3	4	12	NO	0.200	0.167	0.03333
14	3.5	4	14	NO	0.200	0.167	0.03333
15	4	4	16	NO	0.200	0.167	0.03333
16	2	6	12	NO	0.200	0.167	0.03333
17	2.5	6	15	NO	0.200	0.167	0.03333
18	3	6	18	NO	0.200	0.167	0.03333
19	3.5	6	21	NO	0.200	0.167	0.03333
20	4	6	24	NO	0.200	0.167	0.03333
21	2	8	16	NO	0.200	0.167	0.03333
22	2.5	8	20	NO	0.200	0.167	0.03333
23	3	8	24	NO	0.200	0.167	0.03333
24	3.5	8	28	YES	0.200	0.167	0.03333
25	4	8	32	YES	0.200	0.167	0.03333
26	2	10	20	NO	0.200	0.167	0.03333
27	2.5	10	25	NO	0.200	0.167	0.03333
28	3	10	30	YES	0.200	0.167	0.03333
29	3.5	10	35	YES	0.200	0.167	0.03333
30	4	10	40	YES	0.200	0.167	0.03333

P(ALERT) = Sum of pT for all ALERT solutions
= 5 x .03333 = .1667 = 16.7 percent

The P(ALERT) is the same as was previously obtained without using the probability columns.

Now to illustrate the use of non-uniform distributions, the example is repeated with normal distributions for V1 and V2, which are approximated with triangular distributions. The probabilities pV1 and pV2 of variables V1 and V2 are estimated (see the previous section's discussion of triangular distributions):

LIMM=0 LIMX=25 SOLUTION=V1*V2, V1 and V2 = Normal Distributions

N	V1	V2	SOLUTION	ALERT?	pV1	pV2	pT
1	2	0	0	NO	0.067	0.049	0.003
2	2.5	0	0	NO	0.233	0.049	0.011

3	3	0	0	NO	0.400	0.049	0.020
4	3.5	0	0	NO	0.233	0.049	0.011
5	4	0	0	NO	0.067	0.049	0.003
6	2	2	4	NO	0.067	0.167	0.011
7	2.5	2	5	NO	0.233	0.167	0.039
8	3	2	6	NO	0.400	0.167	0.067
9	3.5	2	7	NO	0.233	0.167	0.039
10	4	2	8	NO	0.067	0.167	0.011
11	2	4	8	NO	0.067	0.284	0.019
12	2.5	4	10	NO	0.233	0.284	0.066
13	3	4	12	NO	0.400	0.284	0.114
14	3.5	4	14	NO	0.233	0.284	0.066
15	4	4	16	NO	0.067	0.284	0.019
16	2	6	12	NO	0.067	0.284	0.019
17	2.5	6	15	NO	0.233	0.284	0.066
18	3	6	18	NO	0.400	0.284	0.114
19	3.5	6	21	NO	0.233	0.284	0.066
20	4	6	24	NO	0.067	0.284	0.019
21	2	8	16	NO	0.067	0.167	0.011
22	2.5	8	20	NO	0.233	0.167	0.039
23	3	8	24	NO	0.400	0.167	0.067
24	3.5	8	28	YES	0.233	0.167	0.039
25	4	8	32	YES	0.067	0.167	0.011
26	2	10	20	NO	0.067	0.049	0.003
27	2.5	10	25	NO	0.233	0.049	0.011
28	3	10	30	YES	0.400	0.049	0.020
29	3.5	10	35	YES	0.233	0.049	0.011
30	4	10	40	YES	0.067	0.049	0.003

P(ALERT) = Sum of pT for all ALERT solutions
= .084 = 8.4 percent

For this example, a lower probability or error occurs (8.4%) when assuming triangular (pseudo-normal) distributions for the variables, compared to the probability obtained (16.7%) when assuming uniform distributions.

The Design Analysis Handbook

2.7.4 Bounding the Probability of an ALERT

Because the probability distributions of many variables are generally either unknown or very hard to obtain, an estimating procedure can be used to bracket the probability of an ALERT. The procedure requires that the probability be computed twice. One calculation assumes all variables have uniform distributions, which usually yields a maximum (pessimistic) probability estimate. The other calculation assumes all variables have triangular (pseudo truncated normal) distributions, which usually yields a minimum (optimistic) probability. The resultant min/max range of ALERT probabilities can be used for risk assessment.

THE ANALYST'S RULE FOR ESTIMATING
THE MIN/MAX PROBABILITY OF AN ALERT
Assume that all variables have uniform distributions and calculate the MAX (worst case) probability. Next, assume that all variables have triangular distributions and calculate the MIN (best case) probability.

To indicate the potential severity of the ALERT, the maximum probability can be listed under the ALERT status in the Analysis Table:

CASE/DESC	UNITS	SPEC Minimum	SPEC Maximum	ACTUAL Minimum	ACTUAL Average	ACTUAL Maximum	STATUS
1.1 DC Gain	V/V	185	215	183.11	196.18	209.95	ALERT 0.304%

2.7.5 Why a Monte Carlo Analysis Is Not Optimum for Risk Assessment

A commonly used method of probability analysis is the Monte Carlo simulation. Monte Carlo techniques employ a dart-board approach to design checking, (i.e., try the design a bunch of times with pseudo-randomly selected part variations and see what happens). After a large number of simulations, a set of results is obtained which provides a sampling of the expected range of performance. Although these data may be very useful for design centering or for gross yield estimating, they are not guaranteed to identify worst case performance boundaries.

In addition, quality requirements generally demand that probabilities of design failure be fractions of a percent or even much less. For such small failure rates it is unlikely that the Monte Carlo sampling would indicate the likelihood of meeting specification unless you ran an enormous number of simulations, which is generally not practical from a cost or time standpoint. With Monte Carlo, since you are using pseudo-random samples, the best you will be able to conclude is something like "if I were to repeat this Monte Carlo analysis of 10,000 simulations many times, on average 95% of my yield estimates would be accurate."

By contrast, enhanced Worst Case Analysis is a deterministic approach which actively explores the total volume of design outcomes, probing its interior and nooks and crannies to identify any out-of-spec results. This positive survey of the entire design range identifies the minimum/maximum performance limits, the performance average, and the sensitivities of all design parameters. By dividing the volume of detected out-of-spec conditions by the total volume of design outcomes, one obtains the probability of occurrence of the out-of-spec conditions.

For risk assessment, therefore, the preferred approach is to first perform a worst case analysis, and then use the worst case results to perform a probability analysis. With this approach worst case performance limits are clearly identified, and a conservative estimate of the probability of unacceptable results is obtained.

To summarize the differences between the two methods, Monte Carlo is a form of testing, where numerous simulated "prototypes" are checked, whereas WCA-plus-probabilities is a form of analysis, where a single min/max "prototype" is evaluated, followed by a statistical evaluation of detected out-of-spec cases. These differences can be significant. For example, which analysis approach would you prefer to have been used in the design of the aircraft you fly on?...a Monte Carlo simulation which determined "likely" performance, but which may not have identified the limits of performance of the aircraft; or a worst case analysis which did identify performance limits, and conservatively estimated the probability of occurrence of undesirable results.

To illustrate the differences between Monte Carlo and WCA+ methods, consider the results of a Monte Carlo analysis described in "Use Spice to Analyze Component Variations in Circuit Designs," *EDN*, April 29, 1993:

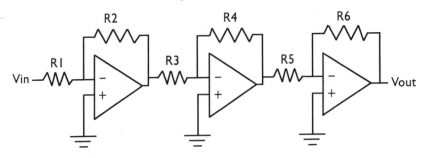

Figure 2.7.5-1
3-Stage Amplifier Used for Monte Carlo Analysis

The PSpice-based Monte Carlo analysis assumed uniform distributions for all resistors, and consisted of 250 trials. The results showed that the amplifier gain varied more than +/-3% after 7 simulations, and by up to +3.9% after 250 simulations.

The results give the designer a feel for the effects of component varia-tions, but what can be concluded? Can the designer state that the amplifier gain error will never exceed +/-3.9%? (No.) Can the designer state that the gain will be within +/-3% some fraction of the time based upon the 7 of 250 simulations? (No.) How many simulations are required to determine whether or not the amplifier yield will be acceptable, given some specifica-tion limits? (Unknown.)

Now let's look at the same circuit using the WCA+ approach. The amplifier gain magnitude is defined by the following expression:

R2/R1 x R4/R3 x R6/R5

The range of values of R1 through R6 are then defined as 990 to 1010 ohms for each resistor. The probability distributions of R1 through R6 are specified as uni-form (other distributions could of course also be defined). Let's assume that the desired gain is 1.0 +/- 3%, or the required gain limits are 0.97 to 1.03.

The gain is calculated, resulting in (a) worst case gain values of 0.9418 to 1.0618 (-5.82% to +6.18% error), and (b) an estimated probability of exceeding specification of 1.56%. Using the calculation data, a probability plot for the equation is provided below:

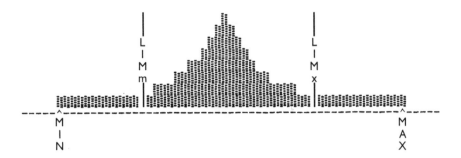

Figure 2.7.5-2
WCA-Generated Probability Plot of 3-Stage Amplifier

As another example, for a low pass filter design you could use a simulator with Monte Carlo processing and run a bunch of curves of the filter response to obtain a sampling of the range of possible performance. But would this be adequate to validate your design? Probably not. But by using worst case analysis methodology, the limits of performance can be identified, along with the probability of error (see graph and plot below):

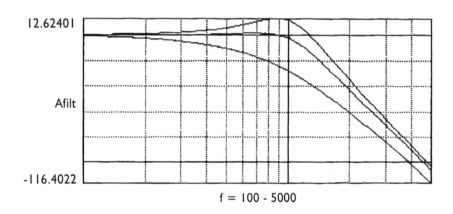

$f = 100 - 5000$

Figure 2.7.5-3
WCA-Generated Frequency Response Envelope
for 8-Pole Low Pass Filter

– 57 –

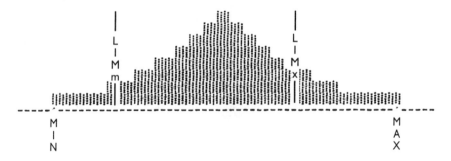

Figure 2.7.5-4
WCA-Generated Probability Distribution of
800 Hz Response of 8-Pole Low Pass Filter

The filter specification requires an attenuation of -15/+3 dB at the 800 Hz cutoff frequency. The analysis shows that the filter will not meet this requirement approximately 9.5% of the time.

2.7.6 Interpreting the Results of the Probability Analysis

Before passing judgement on the results of a probability calculation, additional consideration should be given to the nature of the variables involved in the analysis. This will be demonstrated by example:

Assume a circuit design relies upon a battery voltage as an input. The battery voltage can range from 11 to 13.8 volts, and the designer assumes a uniform distribution to reflect a linear discharge cycle. A calculation of the probability distribution of the circuit's performance determines that the circuit will fail 0.01% of the time as a function of battery voltage; the influence of other variables is nil. The designer tells his boss that the circuit has a 0.01% failure rate, the boss says that sounds acceptable, and a hundred units of the related product are built and shipped the first month.

On the second month, after all the units have failed and have been returned by irate customers, while reading the "help wanted" ads the designer realizes his mistake: the 0.01% failure rate was for each discharge cycle of the battery, not for the product lifetime . Since the battery would typically undergo a charge/discharge cycle one or two times a month, a product failure would occur each month! In fact, even if the error only occurred for a teeny tiny sliver of voltage which represented only 0.000005% of the range of battery voltages, since that voltage would be hit once a cycle, the error would still occur monthly; i.e., any error rate per battery cycle still translates to a 100% failure rate for the product the first month.

The moral is that a definite distinction exists between variables whose values are relatively static after their initial selection, versus variables whose values can vary or cycle during the lifetime of the product. Examples of the former are component initial tolerances and temperature coefficients; e.g., if you pick a resistor from a parts bin, its initial tolerance and temperature coefficient won't change thereafter, and the related distributions are relevant to the product lifetime. Examples of the latter are temperature, AC line voltage, and dynamic loads, where in general it is expected that variations will occur often during a product's lifetime. Therefore, to properly evaluate a probability calculation, the following rule should be followed:

THE ANALYST'S RULE FOR EVALUATING PROBABILITY RESULTS
If a variable can vary over its min/max range within a period which is less than the product lifetime, then the probability results should be examined or recalculated for representative fixed values of the variable across the variable's range.

For example, if the probability results are heavily dependent upon temperature (this can be determined by the sensitivity analysis; see Section 2.10), then it would be wise to examine the effects of temperature at various operating points. It is possible that a small value of P(ALERT) obtained by considering the full distribution of temperature may jump to a very high and unacceptable value if temperature is evaluated near one of its operating extremes.

As can now be appreciated, the casual interpretation of probability results can lead to significant errors. This is also true for other averaging techniques, such as root sum square, where sometimes designers incorrectly toss all the variables into the calculation soup without regard for their periodicity; this will *not* yield an award-winning recipe. On the contrary, the designer/analyst is once again forced to *think*, to consider the ramifications of the design variables.

In some cases, a given variable must be evaluated depending upon the specific application. Consider the design of a power supply, where it is desired to define the required output power. To avoid overdesign, the power supply designer can take advantage of the distribution of the sum of individual IC operating currents, provided that each IC's current, once determined by initial selection of each individual IC, remains relatively static. For a dynamic load, however, this statistical advantage is reduced or eliminated, since the current will change often during the lifetime of the power supply. Therefore the performance at load extremes should be examined. Likewise, a distribution of IC currents which spans different IC operating modes (e.g., standby and operate) would generally be inappropriate; a better choice would be a distribution centered around the mode with highest current.

2.8
RECONCILIATION AND FEEDBACK

WCA results are compared to test and simulation results, with any disparities reconciled by a review process. Also, it is desirable to provide ALERT results to the design team on a real-time basis whenever possible. After analysis results are confirmed, ALERT cases can be evaluated by a risk assessment process:

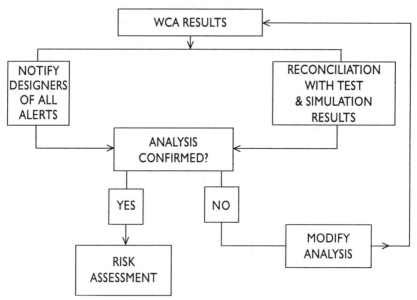

Figure 2.8-1
WCA Results Reconciliation Flow Diagram

2.9
PERFORM A RISK ASSESSMENT

When an out-of-specification (ALERT) condition is identified, engineering and/or business managers must determine whether to fix the ALERT, or accept the risk of leaving the potential problem in place. The general procedure summarized below uses the probability data generated by WCA+, and provides the project manager with a rational basis for risk assessment:

THE MANAGER'S GUIDE TO RISK ASSESSMENT

STEP 1: Perform a Worst Case Analysis

STEP 2: Determine the system consequence of each ALERT. If unacceptable, FIX, else

STEP 3: Determine the probability of occurrence of the ALERT. If not negligible, then

STEP 4: Determine the cost of fixing the ALERT later (warranty repair cost times the total production quantity).

STEP 5: Determine the cost of fixing the ALERT now (nonrecurring redesign cost, plus recurring unit alteration cost times the total production quantity).

STEP 6: Multiply the probability of occurrence (Step 3) times the cost to fix later (Step 4) to arrive at the factored ALERT cost.

STEP 7: If the factored ALERT cost (Step 6) is greater than the cost to fix now (Step 5) then FIX.

Establishing Design Quality Goals

How reliable does a design have to be? This depends on the overall quality which your company has established for its product *design* (as distinct from workmanship, component, and other quality yardsticks), stated

The Design Analysis Handbook

in terms of allowable design failure rate. Based upon a total allowable design failure rate Ftot, a lower-level failure rate allowance can be assigned to each design parameter by using the following formula:

Fdp = Ftot/(Pf+Ps) * 1/FF = design parameter failure rate

where Ftot = total design failure rate allowance
      Pf = total number of functional parameters
      Ps = total number of stress parameters
      FF = failure factor = estimated maximum portion of cases
              which will have failure rates greater than zero; i.e., FF
              indicates the design team's confidence in the initial design.

The design parameter failure rate Fdp can be used during the course of the design and related worst case analysis to monitor design quality.

EXAMPLE: Ftot = 1 design failure allowed per ten thousand systems
           = 0.0001

Pf = 20 (20 system performance specifications)

Ps = 35 x 3 = 105 (35 components times estimated 3 stress parameters per component)

FF = 0.01; i.e., assume 1 percent of cases will have failure rates greater than zero

Fdp = 0.0001/(20+105) * 1/0.01 = 0.00008, or 0.008 percent

During the worst case analysis, the probability of failure (Pfail) is calculated for each design parameter which exceeds specification. Any single Pfail which exceeds Fdp is categorized as an ALERT case and is subjected to risk assessment. Cases with probabilities of failure greater than zero but less than Fdp are categorized as CAUTION cases. ALERT cases which are determined to be significant receive immediate corrective action. An exception may be made for ALERT cases which are difficult or expensive to correct; such cases can be held in abeyance until the analysis is complete.

Following completion of the analysis, the Pfail probabilities are summed to determine the total actual design failure rate. If the sum is less than Ftot, then no further design corrections are necessary. Otherwise, the higher-risk cases receive corrective action until the sum of all Pfail is less than Ftot.

2.10
CORRECT AND OPTIMIZE THE DESIGN

If the analysis detects an ALERT, corrective action and design optimization are facilitated by an understanding of the equation's sensitivities. Sensitivities identify the variables which have the greatest relative effect on a given equation. The classic definition of sensitivity is

$$\text{SENSITIVITY}= \frac{\text{\% Change in Result due to change in Variable V}}{\text{\% Change in Variable V}}$$

Using the classical definition, sensitivities are often calculated by varying each variable a small amount, say 1%, and then observing the effect on the output. For example, a 1% change in resistor value which results in a 15% change in output yields a sensitivity of 15 for the resistor. However, such traditional sensitivity calculations can be very misleading because *the fixed factor applied to each variable does not in general represent the variation to be expected in a given application.* For example, calculations using 1% resistor change, 1% supply voltage change, and 1% temperature change yield classical sensitivities whose significance is distorted and obscured by the fact that (for example) the actual resistor change will be 5 times higher, the actual supply voltage change will be 10 times higher, and the actual temperature change will be 400 times higher!

Therefore it is much better to vary each variable over the *expected* range of the variable for the application, and examine the resultant change to the output. These results can be normalized to equal 100 percent, which provides a table of sensitivity ratios which indicate at-a-glance which variable has the most significant effect on the output.

For example, assume that an equation for temperature yields a min/max result of 30 to 70 degrees F versus several variables. Assume that one of the variables is velocity, and velocity varies from 200 to 300 fps. Further assume that temperature varies from 35 to 50 degrees F versus the change in velocity from 200 to 300 fps. The sensitivity of temperature to velocity is

SENS(temp) = % change in temperature due to velocity /
             % change in velocity
         = (50-35)/(70-30) / (300-200)/(300-200)
         = 15/40 / 1
         = 0.375
         = 37.5 percent

Therefore, 37.5% of the change in temperature is due to the change in velocity, which indicates that in this example temperature is heavily dependent on velocity.

Using the example from section 2.5.3, the normalized sensitivities for Adc are:

Ta	.82%
TcA	.18%
TiR1	-35.96%
TcR1	-13.54%
TiR2	35.96%
TcR2	13.54%

In the referenced example, the ALERT occurs at the lower range of the solutions; i.e., the minimum actual value of Adc (Adc = 183.11) is less than the required minimum specified value (LIMM = 185). The probability plot of the solutions provides a very helpful visualization of the situation. Using the sensitivities table, the design center can be shifted in the desired direction (higher) by adjusting a selected variable by the Design Correction Factor (DCF):

$$DCF = 1 + \frac{(Vmax-Vmin)*(SavgNEW-Savg)}{Vavg*Vsens*(Smax-Smin)}$$

where   Vmax = maximum value of variable
           Vmin = minimum value of variable
           Vavg = average value of variable
           Vsens = sensitivity of variable (stated as fraction)

SavgNEW = desired average value of solutions
           = (LIMM+LIMX)/2

           where LIMM = minimum spec limit
                  LIMX = maximum spec limit

Smax = maximum solution
Smin = minimum solution
Savg = average solution
      = sum of all solutions / number of solutions

To optimize the design, the selected variable is multiplied by the DCF. If it is more convenient to apply an adjustment offset rather than a factor, or if the selected variable has a zero average (+/- tolerance around zero), then the Design Correction Amount should be computed:

$$DCA = \frac{(Vmax-Vmin)*(SavgNEW-Savg)}{Vsens*(Smax-Smin)}$$

To optimize the design, add the DCA to the selected variable.

For the previous example, assume that it is convenient to adjust variable TiR1:

$$Vmax = 1.02$$
$$Vmin = 0.98$$
$$SavgNEW = (215+185)/2 = 200$$
$$Savg = 12555.38/64 = 196.18$$
$$Smax = 209.95$$
$$Smin = 183.11$$
$$Vsens = -0.3596$$
$$Vavg = 1.0$$

DCF = 1+(1.02-.98)*(200-196.18)/(1.0*-0.3596*(209.95-183.11))
     = 0.9842

By substituting TiR1*DCF for TiR1, the new analysis results are:

CASE/DESC	UNITS	——SPEC——		———ACTUAL———			STATUS
		Minimum	Maximum	Minimum	Average	Maximum	
1.1A DC Gain	V/V	185	215	185.93	199.26	213.31	OK

Successful centering is confirmed by the new average value of 199.26, which is very close to the desired design center of 200. Successful design centering is also confirmed by the new probability distribution plot:

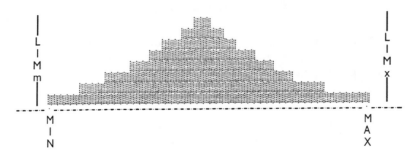

Figure 2.10-1
Probability Plot of Design-Centered Amplifier

CHAPTER THREE

DESIGN VALIDATION TOPICS AND TIPS

3.1
SCREENING THE ANALYSIS

For worst case purposes, it is not necessary to compute the precise result for each of the possible combinations of parameters; it is only necessary to obtain fairly precise values for the limits of performance. In fact, by using simplifying but conservative assumptions, it is often fairly easy to obtain limit values which are sure to lie on or slightly outside the true worst case performance boundaries for a set of analysis cases. Although such results are somewhat pessimistic (a little worse than worse case), the savings in analysis time can be substantial. This leads to the analyst's simplification rule:

THE ANALYST'S SIMPLIFICATION RULE
To start, use simple and conservative min/max boundary equations.

If no ALERT is detected, substantial time has been saved with no penalty, and by inductive reasoning a whole class of analysis cases has been proven OK.

This approach is frequently employed on Stress Margins analyses, where it is possible to assign a basic set of constraints to a large group of components. For example, if a circuit has a group of 1/8 watt resistors and the maximum available supply voltage for the group is +5.25V, then you can calculate the minimum resistance value which will not be overstressed, assuming the entire 5.25 volts is applied to the resistor:

$$Rmin = (Max\ volts)^2/(Rated\ watts)$$

$$= 5.25^2/0.125$$
$$= 27.56/0.125$$
$$= 220.5 \text{ ohms}$$

After solving for the minimum value, you have proven that all resistors with values greater than 220.5 ohms are OK, and do not need individual analysis. However, it is not necessarily true that all resistors with values less than the minimum value are an ALERT, since a given resistor may have less than 5.25V applied; i.e., resistors less than 220.5 ohms need to be analyzed individually. Also, you must be very careful that your underlying assumption (5.25V max across any resistor) is true. In this example, one needs to check to be sure that no reactive components or transformers are present which can generate voltages in excess of the supply voltage.

3.2
COMPUTER-AIDED WCA

Worst Case Analysis can be automated using spreadsheets, or by adapting math-based software packages to WCA methodology. Of the numerous available programs, only DACI's *Design Master*™ is dedicated to Worst Case Analysis. *The Design Master* embodies the Worst Case Analysis *Plus* methodology described herein, including the probability and sensitivity features required for risk assessment and design optimization.

Also, in-house custom software programs can be developed. Algorithms can be employed which identify worst case values and probabilities from sample sets of calculations, thereby increasing analysis efficiency. Some worst case calculation algorithms require the generation of sensitivities, which can also be used for design correction and optimization (see 2.10).

3.3
NONLINEAR EQUATIONS

Nonlinear functions include clipping, step, squarewave, pulse, triangular wave, sawtooth wave, staircase, rectification, peak detection, etc. All of these functions can be generated by closed-form equations. For example, although the nonlinear IF-THEN-ELSE function is generally available in spreadsheet or other math software, it is instructive to see how such a function can be created in closed form:

RESULT = 0.5*(ELSE+THEN)+0.5*(ELSE-THEN)*(X-LIMIT)/ AB(X-LIMIT)

where AB = absolute value function

For the above nonlinear equation, the solution (RESULT) will equal THEN if the variable X is less than LIMIT; RESULT will equal ELSE if X is greater than LIMIT; and RESULT will equal the average of ELSE and THEN if X equals LIMIT.

X is the variable of interest, and can be defined as an independent variable or a subfunction. ELSE, THEN, and LIMIT can be defined as constants, independent variables, or subfunctions, which makes the IF-THEN-ELSE equation very powerful.

For a simple IF-THEN-ELSE application, assume that an amplifier's output voltage Vo equals Vi * K, where Vi = input voltage and K = gain. For a gain switching design, assume that gain K = K1 when Vi is less than a pre-defined input threshold of Vlimit volts, and K = K2 when Vi is greater than Vlimit. Stated in IF-THEN-ELSE terms,

IF input Vi is less than Vlimit THEN Vo = Vi*K1 ELSE Vo = Vi*K2

Using the IF-THEN-ELSE equation,

Vo = 0.5*(ELSE+THEN)+0.5*(ELSE-THEN)*(X-LIMIT)/AB(X-LIMIT)

ELSE = Vi*K2

THEN = Vi*K1

LIMIT = Vlimit

X = AB(Vi)   (use absolute value to account for +/- input voltages)

Although terms Vi, K1, K2, and Vlimit are defined as independent variables in this example, they could well have been defined as subfunctions; i.e., very complex IF-THEN-ELSE structures can be created.

For another example, a linear function such as an amplifier output can be clipped to minimum and maximum values by the following clipping function:

$$CLIP(X) = .5*(LO+HI+AB(X-LO)-AB(X-HI))$$

where LO = lower clipping value
HI = upper clipping value
X = variable which ranges from below LO to above HI
AB = absolute value function

For X = SIN(wt), LO = -0.8 to -0.5, and HI = 0.25 to 0.95, the graph of CLIP(X) is shown below:

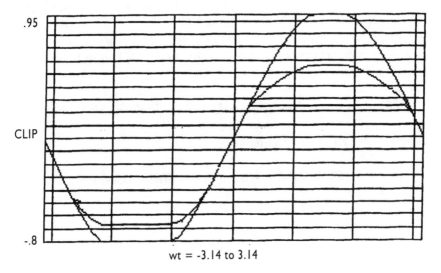

wt = -3.14 to 3.14

Figure 3.3-1
Graph of Clipped Waveform

3.4
THE SOFTWARE ANALYSIS

Software validation methodology at present is in its infancy, and consists mainly of test, test, test; there is no clear counterpart to WCA for software design. However, DACI has found that Single Point Failure Analysis (SPFA) and Fault Tree Analysis (FTA) methodologies (see sections 4.1 and 4.2) can be adapted to meet software safety validation requirements.

The implications of SPFA and FTA for software are the same as they are for hardware: "bottlenecking" is a powerful technique for controlling what may otherwise be an intractable analysis problem. Indeed, some disturbing results concerning software validation have been published (*Science News* Vol. 140, 14 Dec. 1991). A paper by Butler and Finelli of NASA Langley Research Center demonstrates that validation of a complex software program can take years or even decades of testing on the fastest available computers; i.e., validation is not feasible for very complex programs.

A software-SPFA or software-FTA for critical design applications is recommended. The initial steps of one of these analyses will determine whether or not the software complexity is too severe to yield to a full analysis.

3.5
APPLYING TOLERANCES

Tolerance sub-variables are a convenient means of defining the tolerances of higher variables. For example, a resistor variable R1 may have an associated initial tolerance, which can be defined as TOLR1i. If TOLR1i is stated in terms of its actual value, e.g., +/- 10 percent or -0.1 to +0.1, then the complete expression for R1 is

$$R1*(1+TOLR1i)$$

where

NAME	MIN	MAX	DESCRIPTION
TOLR1i	-0.10	0.10	Initial tolerance, +/- 10%

The tolerance can be compacted by changing from a +/- format to the multiplier format, and the complete expression for R1 is

R1*TOLR1i

where

NAME	MIN	MAX	DESCRIPTION
TOLi	0.90	1.10	Initial tolerance, +/- 10%

It is important to use as many tolerance variables as there are independent parameters, even if the tolerances for the parameters are the same. Otherwise, the tolerance values for each parameter would track (be the same value), and worst case combinations could easily be missed. If the variables are dependent, then the same tolerance variable should be used. For example, assume that three variables are used to define the unknown X:

X = VAR1*VAR2/VAR3

If each variable has an initial tolerance of +/- ten percent, and their tolerances are not related to each other, then the correct equation is

X = VAR1*TOLi1*VAR2*TOLi2/(VAR3*TOLi3)

where

NAME	MIN	MAX	DESCRIPTION
TOLi1	0.90	1.10	Initial tolerance, +/- 10%
TOLi2	0.90	1.10	Initial tolerance, +/- 10%
TOLi3	0.90	1.10	Initial tolerance, +/- 10%

However, if the initial tolerances of VAR1, VAR2, and VAR3 track (all low, all nominal, all high, etc.), by being related to a common process, then a single tolerance variable should be used:

X = VAR1*TOLi*VAR2*TOLi/(VAR3*TOLi)
  = VAR1*VAR2*TOLi/VAR3

where TOLi is as defined previously.

This latter expression assumes that all variable tolerances track exactly, which is the ideal case. In practice, tracking tolerances should be defined. For example, assume that the initial tolerance for VAR1 is TOLi1, and that the initial tolerances for VAR2 and VAR3 will track VAR1's tolerance within +/- 1.0 percent:

X = VAR1*TOLi1*VAR2*TOLt2/(VAR3*TOLt3)
where

NAME	MIN	MAX	DESCRIPTION
TOLi1	0.90	1.10	Initial tolerance, +/- 10%
TOLt2	0.99	1.01	Tracking tolerance, +/- 1%
TOLt3	0.99	1.01	Tracking tolerance, +/- 1%

Tolerances can be compounded to combine other effects. For example, assume that Y is a function of variable VAR4 which, in addition to an initial tolerance of 3 percent, has a temperature coefficient of +/- 300 ppm/C relative to 25C. This can be expressed as:

Y = VAR4*TOLi4*(1+TC4*(Ta-25))

where

NAME	MIN	MAX	DESCRIPTION
TOLi4	0.97	1.03	Initial tolerance, +/- 3%
TC4	-3.00E-4	3.00E-4	Temperature coef., +/- 300ppm/C
Ta	-25	85	Ambient temperature, C

3.6
ABSOLUTE VS. DRIFT ACCURACY

Absolute accuracy is the accuracy of a system with all variables considered, including initial tolerances. Drift accuracy is the accuracy of a system with all variables considered, but with the effects of initial tolerances reduced by an initial calibration process.

The Design Analysis Handbook

When analyzing for drift accuracy, the *variables for initial tolerances cannot be reduced any lower than the measurement accuracy of the calibration instrumentation*. The prior VAR4 example is shown below, modified for a drift accuracy analysis with an initial calibration accuracy of +/- 0.5%:

$$X = VAR4*TOLd4*(1+TC4*(Ta-25))$$

where

NAME	MIN	MAX	DESCRIPTION
TOLd4	0.995	1.005	Drift tolerance, +/- 0.5%
TC4	-3.00E-4	3.00E-4	Temperature coef., +/- 300ppm/C
Ta	-25	85	Ambient temperature, C

3.7
ENSURING REALISTIC RESULTS

When defining an expression, the user must take care to eliminate all dependent variables, otherwise worse than worst case results may be obtained. For example, assume that an expression for voltage V contains the independent variable Ta for ambient temperature and the dependent variable Vf for a diode's forward voltage; i.e., Vf's value is a function of Ta:

$$V = 0.44*(Ta-25)+Vf$$

If the temperature worst case limits are used (e.g., 0 to 75 C) and the Vf worst case limits are used (e.g., 0.49 to 0.655 volts) without accounting for the dependency of Vf on Ta (e.g., -1.8 to -2.2 millivolts/C) then a result may be obtained which cannot actually occur. Solving for the maximum value of V,

$$V(max) = 0.044*(75-25)+0.655 = 2.855 \text{ volts}$$

In the incorrect solution above, the maximum value of Vf is used when the maximum value of Ta is used, but this cannot occur since Vf has a negative dependency upon Ta. Therefore, Vf should be defined as an equation which shows its dependency on Ta, e.g.,

Vf = Vfi+VfTC*(Tj-25)

where   Vfi = forward voltage at 25 C = 0.6 volts
        VfTC = temperature coefficient of Vf = -1.8 to -2.2 millivolts/C
          Tj = junction temperature of diode, approximately equal to Ta
             for low power = 0 to 75 C

and
V = 0.44*(Ta-25)+Vfi+VfTC*(Ta-25)

V is now properly stated in terms of constants and independent variables; the dependent variable Vf has been eliminated.

Solving for V(max),

V(max) = 0.044*(75-25)+(0.6-1.8E-3*(75-25)) = 2.71 volts

The correct worst case maximum value is therefore 2.71 volts, not 2.855 volts.

In summary, to prevent erroneous "worse than worst case" results, an equation being analyzed should contain only constants and independent variables.

Chapter Four ⎯⎯⎯⎯⎯⎯⎯⎯⎯

SAFETY ANALYSES

After performing a Worst Case Analysis, an additional layer of analysis activity is often required for critical systems. A critical system is a system which, if it fails, may pose a hazard to personnel, equipment, or the environment. Some guidelines for fail-safe design are presented in Table 4-1.

⎯⎯⎯⎯⎯⎯⎯⎯⎯⎯⎯⎯⎯⎯⎯⎯⎯⎯⎯⎯⎯⎯⎯⎯⎯⎯⎯⎯⎯⎯⎯⎯

Table 4-1
GUIDELINES FOR FAIL-SAFE DESIGN

1. GENERATE A SYSTEM-LEVEL SELF TEST PLAN WHICH IDEN-TIFIES THE FOLLOWING:

   a. CRITICAL FAILURES: failures which may result in a significant hazard to personnel, equipment, or the environment.

   b. CRITICAL FUNCTIONS: system functional blocks wherein in-correct operation may result in a Critical Failure.

   c. SELF TEST FUNCTIONS (ALARMS): system blocks/actions which check the performance of Critical Functions.

2. IMPLEMENT THE SELF TEST FUNCTIONS USING THE FOL-LOWING GUIDELINES:

   a. Each Self Test Function should exercise a Critical Function output for both its (simulated) failed and normal states: If the function is not tested in its simulated "failed" state, the Self Test output may fail in the "tests good" condition, which will not be detected. If a subsequent true failure of the Critical Function occurs, it will not be detected. (Checking the good/bad output states of an Alarm is referred to as an Alarm Check.)

b. Check the transition from Test to Normal mode: After checking the output of a Critical Function in the Test mode, check the output in the Normal mode at a different operating point than the Test mode. If this is not done, and the Self Test fails in the "test" position, the test signal may be misinterpreted as a measured signal.

c. Test a dual-polarity Critical Function at both polarities: For a circuit operating from dual power supplies it may be possible for the circuit to function correctly for signals of one polarity, but to have clipping/distortion error for signals of the opposite polarity. Therefore a self test of a single polarity may test "good" while a measurement of the opposite polarity will be in error.

d. Test a Critical Function at its response limits: Some failures limit the response of system to less than the desired output, while not affecting the output response near the center of the range. Therefore a self test of a center value may test "good" while a measurement of a value near the output limits will be in error.

e. Test a dynamic Critical Function with dynamic test signals: Waveform or timing distortion errors which may result in Critical Failures should be checked by using a dynamic test signal to test the Critical Function. A dynamic test can also be used to satisfy the dual-polarity and response limit tests described above.

3. GENERATE A SELF TEST DESIGN REPORT (block diagram and flow diagram format) which describes each Self Test Function and identifies how each is used to satisfy the Self Test Guidelines for each Critical Function.

4. VALIDATE THE DESIGN VIA A SINGLE POINT FAILURE ANALYSIS to determine whether any failure mode of any system block can result in a Critical Failure. Note that failures which result in "no

effect" (undetected failure) should be reviewed to ensure that a subsequent second failure cannot cause a Critical Failure.

The Single Point Failure Analysis should be applied to all critical systems, since it is an exhaustive survey of all potential failure modes and consequences. The Sneak Path Analysis (SPA) should also be applied to critical systems to help identify unintended modes of operation. The distinction is that the SPFA examines improper operation due to component failure; the SPA examines improper operation due to unintended operation but assumes no component failures. The Fault Tree Analysis (FTA) can be additionally applied to well-defined systems wherein the goal is to make sure that the probability of occurrence of defined unacceptable failures is acceptably small.

4.1
THE SINGLE POINT FAILURE ANALYSIS (SPFA)

The Single Point Failure Analysis (also called Failure Modes and Effects Analysis, or FMEA) tabulates every failure mode of every system component and determines the related local and overall system effects. The results are used to identify and eliminate failure modes which have unacceptable consequences.

The SPFA examines the effects of presumed component or assembly defects. It does not address parameter variation (worst case design) considerations, and therefore is not intended to detect design deficiencies, nor will it. It will identify the *consequences* of design deficiencies that result in component failure.

The SPFA procedure is:

1. Define the critical failures; e.g.,

    CRITICAL FAILURES:
    a. Brake system fails.
    b. Overpressure condition in chamber 1.
    c. Undetected failures related to critical functions.
    Etc.

Note that an *undetected failure* which can subsequently result in a critical failure is the same as an initial critical failure.

2. Define the failure modes for each component; e.g.,

COMPONENT TYPE	FAILURE MODES
CAPACITOR	OPEN, SHORT
INTEGRATED CIRCUIT	INPUT OPEN, INPUT SHORT HIGH, INPUT SHORT LOW,OUTPUT OPEN, OUTPUT SHORT HIGH, OUTPUT SHORT LOW
RESISTOR	OPEN, SHORT
SOLENOID	COIL OPEN, SHORT ACTUATOR STUCK IN, STUCK OUT

Etc.

3. Determine the local and system effects for each failure mode; e.g.,

REF	FAILURE MODE	LOCAL EFFECT	SYSTEM EFFECT	STATUS
R1	OPEN	No oscillator output	Lose speed readout	OK
R1	SHORT	No oscillator output	Lose speed readout	OK
C1	OPEN	No oscillator output	Lose speed readout	OK
C1	SHORT	Lose +12V power supply; solenoid S3 deactivates	Can't apply brakes; CRITICAL FAILURE	ALERT

Etc.

If a failure effect is dependent upon operating mode, then an analysis is done for each mode.

A valuable technique for reducing analysis complexity is to group the system being analyzed into small blocks, with a minimum number of interface lines for each block. A key restriction is that all relevant information concerning a given critical function must be transmitted across the selected interface lines; i.e., no critical function should be totally enclosed within a block. By analyzing each possible failure mode at each interface line, by inference the effects of all failure modes of all internal components have been determined. Using this approach, the number of analysis cases can be significantly reduced, with no loss of information regarding ultimate failure effects.

This "bottleneck" approach can be particularly valuable if applied during the initial design of a complex system. Without bottlenecking, some complex systems may prove impractical to validate by analysis.

To facilitate a thorough analysis, the following items should be documented and available: critical failure definitions, safety functions, and self test functions (alarms and alarm checks). If software control is employed, the software functions should be fully documented via flow charts.

4.2
THE FAULT TREE ANALYSIS (FTA)

The Fault Tree Analysis (FTA) identifies a set of unacceptable ("critical") failure effects of particular interest and examines the system to determine the combination of conditions required for the critical failures to occur. If the conditions required for a critical failure have a probability of occurrence which is higher than a defined limit, the design is modified.

The FTA is generally easier to implement (and less expensive) than the SPFA, since the SPFA is an exhaustive survey of an entire system, whereas the FTA examines only defined areas of interest. The FTA is recommended for smaller well-defined designs wherein the goal is to ensure that one or more identified critical failures cannot occur. For example, a defined undesirable result for a detonator is "premature detonation." Starting with "premature detonation" at the top of the tree, the conditions required for false detonation are assembled on the branches beneath:

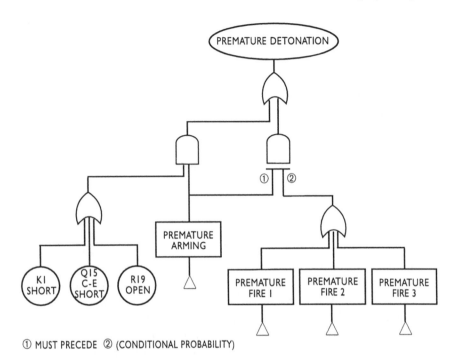

① MUST PRECEDE ② (CONDITIONAL PROBABILITY)

Figure 4.2-1
Example of Fault Tree Diagram

In a good design, the FTA will identify only a small set of conditions which can cause a critical failure, and the probability of occurrence of the conditions will be extremely small.

4.3
THE SNEAK PATH ANALYSIS (SPA)

The Sneak Path Analysis (SPA) identifies unintended system actions which are created by "hidden" paths or events within the system, with the system otherwise functioning normally; i.e., no component failures. Sneak effects which result in unacceptable consequences are eliminated by redesign.

The general procedure for SPA is as follows:

1. Organize the system into small functional blocks. For an electronics

system, for example, this would consist of breaking the system into its basic circuit blocks.

2. Arrange the components of each block into basic topological networks (topographs) which show how energy is transferred. For an electronic system, each topograph would identify a circuit branch.

3. Examine each topograph for unintended energy transfer. For the electronic system, this would be unintended current flow.

4. If unintended energy transfer can occur, assess its consequences.

For example, assume that a parasitic diode within an integrated circuit will allow current to flow up through the IC, even though this is not a normal path. A sneak path analysis would examine the "sneak" path and determine if the path could be activated, and if so assess its consequences. For another example, assume two mechanical levers are normally operated independently. A sneak path analysis would examine whether the levers could be operated simultaneously, and if so assess the consequences.

A portion of a Sneak Path Circuit Analysis is shown below:

TOPOGRAPH 1A   RESULTS: OK

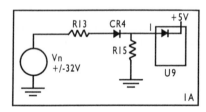

With +5V OFF and Vn = positive, the current into U9-1 through substrate diode is limited by R13. With Vn = negative, current from U9-1 is prevented by blocking diode CR4.

Figure 4.3-1
Portion of Sneak Path Analysis

CHAPTER FIVE ⎯⎯⎯⎯⎯⎯⎯⎯

BAD SCIENCE AND OTHER HAZARDS

5.1
THE RELIABILITY PREDICTION

The Reliability Prediction estimates the mean operating time of a system between failures (MTBF). MTBFs are computed using component normal failure (wear-out) rates. These rates are derived from failure mechanism math models, or are obtained directly from empirically-based failure rate tables.

Many components have wear-out mechanisms which are fairly well defined; e.g, tire tread wear, motor operating hours, light bulb lifetimes, etc. For systems using such well-behaved components, a reliability prediction can be used for budgeting and scheduling maintenance activities. Unfortunately, failure prediction methodology has been inappropriately applied to complex and rapidly evolving systems.

For example, a handbook (MIL-HDBK-217, "Reliability Prediction of Electronic Equipment," prepared by Rome Air Development Center, U.S. Air Force) is used extensively in the electronics industry by both military and industrial engineers for reliability predictions. Investigators have shown that reliance on the methodology and databases of MIL-HDBK-217 may result in less reliability and higher cost, an unpleasant result. The reader is urged to review the articles of Mr. Charles T. Leonard of Boeing Commercial Airplane Division, and Dr. Michael Pecht of the University of Maryland, which critique the claims made by proponents of MIL-HDBK-217. Articles include:

(1) "How Failure Prediction Methodology Affects Electronic Equipment Design," *Quality and Reliability Engineering International*, October 1990.

(2) "Failure Prediction Methodology Calculations Can Mislead: Use Them Wisely, Not Blindly," *NAECON, Failure Predictions Methodology*, March 1989.

(3) "Improved Techniques for Cost Effective Electronics," *IEEE Reliability & Maintainability Symposium*, January 1991.

Two of the key findings from the referenced articles are summarized below:

1. MIL-HDBK-217 methodologies and databases do not result in accurate mean-time-between-failures (MTBF) estimates. MTBFs calculated using the handbook do not track actual results in absolute terms or in trends. This latter point is important, because some people incorrectly think that although the MTBF values generated by the handbook may be too low, they are a relative indicator. In reference (1), an example is presented where a single vendor manufactured several different products for use on the same airplane. The actual MTBFs were compared to the MTBFs predicted by MIL-HDBK-217. The ratio of actual-to-predicted MTBFs varied over a range of 0.22 to 12.2; i.e., there was no correlation between predicted and actual MTBFs.

2. The implications of MIL-HDBK-217 may result in counterproductive designs.

   For example, a major implication of MIL-HDBK-217 is that junction temperature is a key semiconductor device reliability factor, and that operating semiconductors at lower temperatures tremendously increases their reliability. Evidence presented in the articles referenced above suggests that this well-known "rule-of-thumb" has no basis in fact, and that overall system reliability in many cases may be improved by allowing devices to run hotter and eliminating expensive and complex cooling systems.

Others have leveled criticisms against MIL-HDBK-217: see "Critics Put the Heat on MIL-[HDBK]-217" in the February 1991 issue of *Military & Aerospace Electronics* (front page), plus a variety of articles in the *IEEE*

Transactions on Reliability and elsewhere.

MIL-HDBK-217 appears to have been dealt a fatal blow by the trusty hammer of scientific inquiry, wielded by responsible and fair-minded investigators. It also appears that reasonable minds within the higher levels of government are now seriously evaluating the accumulated evidence. Although it will probably take a few years for the dust to settle due to the institutionalizing of MIL-HDBK-217, it is probable that its days are numbered as a defacto government standard. It is also anticipated that the private sector will respond much more quickly, and will discontinue using the handbook except to meet contractual requirements.

As evidence of the eventual demise of MIL-HDBK-217, the U.S. Army has taken steps to discontinue the use of the handbook, directing a shift to the "physics-of-failure" approach to reliability assessment. (This was accompanied by an outbreak of dancing in the streets at several electronics firms.) The new methodology was proposed in a paper by P. Lall and M. Pecht, "An Integrated Physics-of-Failure Approach to Reliability Assessment," Proceedings of the 1993 ASME International Electronics Packaging Conference, vol 4-1, 1993. A related recommended article is "Comparison of Electronics-Reliability Assessment Approaches," *IEEE Transactions on Reliability*, December 1993, by Cushing, Mortin, Stadterman, and Malhotra. The article compares the MIL-HDBK-217 methodology to the physics-of-failure methodology, and provides a clear and concise summary (Table 2 of paper) of the handbook's shortcomings versus the benefits of the new approach.

Q: What if my customer requires a MIL-HDBK-217 reliability calculation?

A: The customer is obviously entitled to whatever they are willing to fund. However, it is recommended that such tasks be isolated from internal company reliability programs, which should not include MIL-HDBK-217-based tasks.

Q: What should my company use in place of MIL-HDBK-217?

A: Reliability efforts should be focused on (a) design validation (worst case analysis plus test/simulation), (b) safety analyses where appropriate (single point failure analyses; fault tree analyses; sneak path analyses), and (3) positive identification, investigation, and corrective action of component, design, process, and workmanship problems. The latter includes the physics-of-failure approach to reliability assessment.

Q: *If I don't use MIL-HDBK-217, how can my company quantify the reliability of its electronic products?*

A: Many of our international competitors do *not* use reliability predictions; they concentrate on building the most reliable products they can. In other words, it's much more important to *have* a reliable product than it is to *claim* to have a reliable product; i.e., the electronics reliability prediction paradigm is obsolete. Electronic product failures (not including misapplication of the product) will be primarily due to process, workmanship, and design problems. By controlling these, your company can develop its own database with which to quantify the reliability of its products. You can supplement your internal reliability database with science-based information developed by sources such as the CALCE Electronic Packaging Research Center at the University of Maryland (301-405-5323).

We should note that at its inception, the handbook was a worthwhile effort to address electronics component reliability in a reasoned manner. But as years passed and the handbook became irrelevant to its original purpose, and even damaging, why did the Air Force's technical experts not advise their managers of the developing weaknesses in the handbook? Why did the managers take so long to listen to the outside critics? What happened to the military's scientific peer review process, which should have afforded the critics a fair and speedy hearing? And finally, why did some of the defenders of the handbook use invective and intimidation as a substitute for reasoned discourse?

(As an example of the latter, the author independently asked Rome Air Development Center for data which supported the handbook, with no response. An angry phone call was subsequently received from a prior employee of RADC, now working for an RADC-sponsored research institute. The caller stated his strong disagreement with the handbook's critics. The author offered to review and publish in the DACI Newsletter a summary of any data the caller was willing to provide which supported the methodology or failure data used in MIL-HDBK-217, since the author had not been able to obtain such data from RADC. The caller dismissed this offer, and, sad to say, engaged in an unprofessional tirade, culminating in the statement that DACI would "be out of business" if we persisted in our criticism of the handbook.)

Maybe this will all be explained in a best-selling book some day. Until then, however, may we be among the first to congratulate the critics of the handbook who have finally prevailed intellectually. By consistently and patiently applying the rules of evidence, they have whittled away the handbook's scientific facade, exposing its true foundation to be an unflattering mix of ego and politics.

As a final note, the Air Force continues to stubbornly defend the handbook. The latest edition as of this writing (version F) even contains deficiencies which Air Force-funded subcontractors recommended be eliminated. Sadly, the author continues to receive reports of the Air Force's ongoing attempts to silence the handbook's critics through intimidation.

5.2
THE TAGUCHI METHOD

"We are not interested in any actual results."
Genichi Taguchi

For other excellent reasons not to use the "Taguchi Method" of design optimization, see the illuminating article by Bob Pease, "What's All This Taguchi Stuff, Anyhow," in the Pease Porridge column of the June 10, 1993 issue of *Electronic Design*.

5.3
SWITCHMODE POWER SUPPLY
CURRENT-MODE CONTROL

We'll be the first to commend vendors for their support of their ICs and other electronic components and products via application notes, seminars, and other forms of engineering assistance. But our commendations are tempered by the fact that such efforts are also a marketing tool which sometimes hype a product to the detriment of the customer, and ultimately the vendor. Current-mode control ICs for switchmode supplies are a case in point.

Over the past several years current-mode controllers have gained wide usage, replacing the traditional voltage-mode controller in many applications. When these ICs first were introduced (1983), they were promoted as having multiple advantages and few (if any) disadvantages. As the years progressed, experience with these ICs caused the perceived advantages to shrink while the disadvantages grew. But each discovered disadvantage was quickly claimed to be minor and easily corrected (with few extra parts, of course) *after* the damage had already been inflicted on the designer who accepted the vendor's original claims.

For example, one of the major claims for current-mode operation was that the current loop part of the topology made the current-mode ICs very easy to compensate (by eliminating the inductive effects of the output filter). Of course, a properly-designed voltage-mode controller is quite stable, but since compensation may be a tad more difficult, why not use a current-mode IC? Many did, only to find that the darn things oscillated! "Not to worry," said the vendor application engineers, "this is due to such-and-such and can be easily corrected if you just add this extra *slope compensation circuit*." "Extra compensation?" the designers thought; "wasn't this IC supposed to be *easier* to compensate?"

But after tacking on a few extra parts, it was found that operation was quite stable...or was it? Additional research (R. D. Middlebrook and others) determined that even with added slope compensation, the basic loop could still be unstable under certain non-trivial conditions. "What?" the designers said. "Maybe this explains some of those mysterious field failures we've been experiencing with our products!"

Additional discoveries were made, such as that a common topology (multiple output buck) required certain precautions with regard to the output filter's inductor design. "What?" the designers said. "Maybe this explains why I wasn't able to get the prototype to work right and I got fired!"

Several other peculiar disadvantages of current-mode controllers have been discovered and studied over the years, making current-mode operation very interesting indeed, as documented in journals and books and presented at seminars. By tapping this knowledge, a designer today can make a reasonably good judgement of whether or not to use current-mode control for a given application. But those engineering pioneers who jumped at using this over-hyped new product in the early years learned a valuable old lesson the hard way:

THE DESIGN ENGINEER'S GUIDE TO USING NEW PRODUCTS
Don't be the first to stick your toe in the water
unless you've got a toe to spare.

5.4
FUZZY LOGIC

Its proponents make you think that it's the greatest discovery since Maxwell's equations, but the truth is probably much more modest. In fact, Fuzzy Logic may just be some old techniques dressed up in new clothes.

As sometimes happens, a branch of technology will rediscover some old truths that have been well known, developed, and refined in another technical discipline. Papers are written, seminars are held, and there is joy in engineerland. Sooner or later, however, all of the hullabaloo is noted by the practitioners of the neighboring technical community, which offer comments like "we've been doing the same thing for years, but we use a different terminology to describe it." This sparks a controversy which is followed by the wailing and gnashing of teeth for awhile by the more passionate advocates of each side of the issue. Eventually the scientific method forces a convergence of views to a reasonable interpretation of the true value of the new technological "advance."

In the case of Fuzzy Logic, control engineers have been creating fuzzy systems for years, by adding comparators, clippers, slope limiters, and other nonlinear elements to standard control systems. For example, a simple 5-comparator circuit can provide "hot," "warm," "neutral," "cool," and "cold" outputs which can be easily combined for fuzzy control purposes. So it doesn't appear as though Fuzzy Logic is a conceptual breakthrough.

On the other hand, it may be that fuzzy proponents have, by virtue of a new way of looking at nonlinear control, something to offer (at this writing the controversy is still swirling). One thing is known, however: the initial over-hyping of Fuzzy Logic has hindered, rather than helped, its evaluation and potential acceptance. Worse, by overpromising, fuzzy control solutions may unduly appeal to less experienced engineers when an old tried-and-true control approach may be the proper design choice.

This is the problem with hype: it promises universal, quick-and-easy, no-pain solutions. You don't have to study, you don't have to research, you don't have to think. Just buy the latest book, go to the latest seminar, follow these simple rules, and your problems will be solved, plus you'll be a hero to boot. That will be $249.95, please, plus tax. No refunds.

CHAPTER SIX _____

ELECTRONICS ANALYSIS
TIPS AND EQUATIONS

" Those who cannot remember the past are condemned to repeat it ."
George Santayana

" You can pay me now...or pay me later ."
Old TV auto service commercial touting
benefits of preventive maintenance

You can say it eloquently or crassly, but the message is the same: it pays to learn all you can today from others, before you pay the higher price tomorrow of learning from experience. Hence this chapter, which contains tips and equations to help you cope with some of the most common yet misapplied, pesky, or in general tough-to-do-right electronics applications.

6.1
POWER AND THERMAL

As anyone who has worked on a lot of electronics equipment will tell you, components frequently fail by overheating, as evidenced by their discolored remains, and often by a telltale unpleasant odor. (As this is being composed, a computer in one corner of the office is emitting the familiar lightly burnt smell, an unwelcome portent of repairs to come.)

SOME COMMON CIRCUIT TOO-HOT SPOTS

CIRCUIT TYPE	REASON FOR PROBLEM
line rectifier circuits:	
- rectifiers	high rms currents
- transformer secondaries	high rms currents
- filter capacitors	high rms ripple currents
- surge resistor	high rms currents
linear regulators:	inadequate heat sinks
switchmode power supplies:	
- input/output filter caps	high rms ripple currents
- current sense resistors	high rms currents
- snubber resistors	high energy transfer

The high incidence of failure-by-overheating may lead some to think that components are more susceptible to failure because of the heat they must dissipate, and therefore endeavor to keep components cool on this philosophical basis alone. However, the real problem is usually *misapplication* of the component through improper heat management, rather than any inherent thermal weakness in the component itself. In fact, material stresses due to thermal cycling or thermal startup transients (the light bulb always pops when it's turned on) can be much more significant than steady-state thermal dissipation.

THERMAL DESIGN & ANALYSIS TIP # 1
Electronic components do not fail due to overheating because heat is inherently bad; they fail because they are operated at temperatures which exceed their maximum rating.

The tip above does not mean that reduced temperature will not increase the average operating lifetime of a part; it may or may not, depending on the part's materials and fabrication. For making operating lifetime judgements, it's best to use science-based data derived from testing and analysis for the specific part or process. Don't rely on inferences from dubious data sources such as MIL-HDBK-217, which some vendors cite as an authoritative basis for their pronouncements on operating lifetime versus temperature.

6.1.1 Power Dissipation Basics

The error of using thermally underrated components often stems from a lack of understanding of the power content of various nonlinear waveforms, which for resistive dissipation is computed by using voltage and current waveform rms values, not *average* values.

As a reminder, an *average* value is the integral of a periodic function over one period, with the result divided by the period:

AVG = INT(function)/Period

EXAMPLE: What is the average value of a fullwave rectified sinewave voltage with a peak voltage of 162.6 volts?

AVG = INT(function)/Period
  = INT(162.6*SIN(wt))/Pi evaluated for wt = 0 to Pi radians
  = 162.6*(-COS(Pi)+COS(0))/Pi
  = 162.6* 0.637
  = 103.5 volts

The *rms* value is the square root of the integral of the square of a periodic function over one period divided by the period:

RMS = SQR(INT(function^2)/Period)

EXAMPLE: What is the rms value of a fullwave rectified sinewave voltage with a peak voltage of 162.6 volts?

RMS = SQR(INT(function^2)/Period)
  = SQR(INT((162.6*SIN(wt))^2)/Pi) evaluated for wt = 0 to
              Pi radians
  = 162.6*SQR((0.5*wt-0.25*SIN(2*wt))/Pi)
  = 162.6*SQR((0.5*Pi)/Pi)
  = 162.6* 0.707
  = 115.0 volts

The rms (effective) values of current (Ieff) or voltage (Veff) can be used to compute the average power dissipation in a resistive component by the relationships Ieff^2*R or Veff^2/R. Also, for sinusoidal volts and amps, if the phase angle pA is known, the average power can be calculated: Pavg = Ieff*Veff*COS(pA). But these are special cases; rms currents and voltages cannot be multiplied together in general to obtain the average power for reactive or variable loads. In such cases, the more general process for obtaining average power is used, which is valid for any situation: instantaneous volts and amps are multiplied together and averaged in accordance with the definition of the mean (Pavg = integral of instantaneous power over a period, divided by the period).

DON'T MISUSE RMS WHEN CALCULATING POWER
Average power can be computed for fixed resistive loads by using rms volts or rms amps. Average power for reactive or variable loads cannot in general be obtained by multiplying rms amps and rms volts together.

The following tables will be a handy reference for a wide variety of power calculations for fixed resistive loads:

Table 6.1.1-1
RMS VALUES FOR VARIOUS WAVEFORMS

Current (I) values are shown. For voltages, substitute V for I

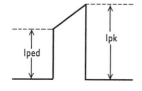

IrRAMP = RAMP+PEDESTAL
= SQR((D/3)*(4*Iped^2+Ipk^2-2*Iped*Ipk))

IrRECT = RECTANGULAR
= SQR(D)*Ipk

IrSINFW = RECTIFIED SINE, FULLWAVE
$$= 0.707*Ipk$$

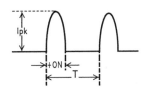

IrSINHW = RECTIFIED SINE, HALFWAVE
$$= 0.5*Ipk$$

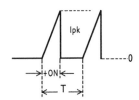

IrSIN = RECTIFIED SINE, PULSED
$$= 0.707*SQR(D)*Ipk$$

$$D = tON/T$$

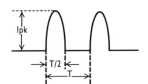

IrSAW = SAWTOOTH
$$= SQR(D/3)*Ipk$$

$$D = tON/T$$

IrTRI = TRIANGLE
$$= SQR(1/12)*Ipp = 0.289*Ipp$$

where SQR = square root function
D = duty cycle
Ipk = peak amps
Iped = pedestal amps
Ipp = peak-peak amps

The Design Analysis Handbook

EXAMPLE: What is the power dissipation in a 0.5 ohm resistor through which a rectangular current is flowing of peak amplitude 3.1 amps and average value 0.25 amps?

$$P = IrRECT^2*R$$
$$= (SQR(D)*Ipk)^2*0.5$$

For a rectangular waveform, the duty cycle D = Iavg/Ipeak

$$= 0.25/3.1 = 0.0806$$
$$P = (SQR(0.081)*3.1)^2*0.5$$

$$= 0.388 \text{ watts}$$

EXAMPLE: What is the power dissipation in the example if it is computed *incorrectly* based upon average current?

$$P(wrong) = Iavg^2*R$$
$$= (0.25)^2*0.5$$
$$= 0.031 \text{ watts!}$$

By incorrectly using the average current, the resistor power is underestimated by a factor of 0.388/0.031 = 12.4 times! (In general, the ratio of the correct power to incorrect power is equal to 1/D, so the smaller the duty cycle, the larger the error).

EXAMPLE: What is the power dissipation in a 1.2 ohm resistor, one side connected to ground, across which a sawtooth voltage exists of base-peak amplitude 3.27 volts and a duty cycle of 40%?

$$P = VrSAW^2/R$$
$$= (SQR(D/3)*Vpk)^2/R$$
$$= (SQR(0.4/3)*3.27)^2/1.2$$
$$= 1.188 \text{ watts}$$

EXAMPLE: What is the power dissipation in a 1.8 ohm resistor through which a triangular current is flowing with a peak-peak amplitude of 1.23 amps and an average value of zero amps (waveform symmetrical around ground)?

$$P = IrTRI^2*R$$
$$= (SQR(1/12)*Ipp)^2*1.8$$
$$= (0.289*1.23)^2*1.8$$
$$= 0.227 \text{ watts}$$

EXAMPLE: What is the power dissipation in the 1.8 ohm resistor from the previous example if the 1.23 peak-peak amp waveform is riding on a DC waveform of 0.36 amps?

The total rms current is the rms combination of the AC and DC components:

$$Iac = SQR(1/12)*Ipp$$
$$= 0.355 \text{ amps}$$

$$Idc = 0.36$$

$$Irms = SQR(Iac^2 + Idc^2)$$
$$= SQR(0.355^2 + 0.360^2)$$
$$= SQR(0.126 + 0.130)$$
$$= 0.506 \text{ amps}$$

$$P = Irms^2*R$$
$$= 0.506^2*1.8$$
$$= 0.461 \text{watts}$$

Table 6.1.1-2
BIPOLAR TRANSISTOR SWITCHING LOSSES, WATTS

PQrRCT = rectangular current waveform (e.g., resistive load)
$$= D*(Ice*VceON+Ibe*Vbe)+0.25*fS*Ice*VceOFF*(tFI+tRI)$$
PQiDM = sawtooth current waveform (e.g., clamped inductive load for discontinuous mode switching power supply)

$$= D*(0.5*Ice*VceON+Ibe*Vbe)+fS*Ice*VceOFF*tFI*Ksf$$

PQiCM = pedestal+ramp current waveform (e.g., clamped inductive load for continuous mode switching power supply)

$$= D*(0.5*(Iped+Ipk)*VceON+Ibe*Vbe)$$
$$+ fS*VceOFF*(Iped*(1+tRR*Kc/tRI)$$
$$* (tRI+tRR*Kc)*Ksr+Ipk*tFI*Ksf)$$

where D = duty cycle
fS = frequency, Hz
Ibe = base-emitter amps
Ice = collector-emitter amps
Iped = collector pedestal amps
Ipk = collector peak amps
Kc = catch diode recovery type: 0, .5, .75 = none, soft, fast
Ksf = turn-off snubber factor: 1 if no snubber, 0 < 1 if snubber
Ksr = turn-on/off snubber factor: 1 if no snubber, 0 < 1 if snubber
tFI = collector current fall time, sec
tRI = collector current rise time, sec
tRR = catch diode reverse recovery time, sec
Vbe = base-emitter volts
VceON = collector-emitter ON volts
VceOFF = collector-emitter OFF volts

EXAMPLE: What is the power dissipation in a power transistor used as the switch element in a discontinuous-mode flyback power supply operating at 30 KHz? The transistor operates at a maximum duty cycle of 40%, a peak current of 2.3 amps, a max saturation voltage of 0.52 volts, a max base voltage of 1.13 volts at a base current of 0.2 amps, a maximum off voltage of 62 volts, a maximum fall time of 500 nanoseconds, and no snubber?

PQiDM = clamped inductive load, sawtooth current waveform

$$= D*(0.5*Ice*VceON+Ibe*Vbe)+fS*Ice*VceOFF*tF1*Ksf$$

$$= 0.4*(0.5*2.3*0.52+0.2*1.13)+30E3*2.3*62*500E-9*1$$

$$= 2.47 \text{ watts}$$

EXAMPLE: What is the power dissipation in the power transistor from the previous example if the supply is operated at 200 KHz?

$$PQiDM = D*(0.5*Ice*VceON+Ibe*Vbe)+fS*Ice*VceOFF*tFl*Ksf$$

$$= 0.4*(0.5*2.3*0.52+0.2*1.13)+200E3*2.3*62*500E-9*1$$

$$= 14.6 \text{ watts}$$

Table 6.1.1-3
FET SWITCHING LOSSES, WATTS

PFrRCT = rectangular current waveform
= $D*Ids^2*rDS+0.25*fS*Ids*VdsOFF*(tFI+tRI)$

PFiDM = sawtooth current waveform
= $D/3*Ids^2*rDS+fS*Ids*VdsOFF*tFI*Ksf$

PFiCM = pedestal+ramp current waveform
= $(D/3)*(4*Iped^2+Ipk^2-2*Iped*Ipk)*rDS$
$+ fS*VdsOFF*(Iped*(1+tRR*Kc/tRI)$
$* (tRI+tRR*Kc)*Ksr+Ipk*tFI*Ksf)$

where D = duty cycle
   fS = frequency, Hz
   Ids = drain amps
   Iped = drain pedestal amps
   Ipk = drain peak amps
   Kc = catch diode recovery type: 0, .5, .75 = none, soft, fast
   Ksf = turn-off snubber factor: 1 if no snubber, 0 < 1 if snubber
   Ksr = turn-on/off snubber factor: 1 if none, 0 < 1 if snubber
   rDS = on resistance, ohms
   tFI = drain current fall time, sec
   tRI = drain current rise time, sec
   tRR = catch diode reverse recovery time, sec
VdsOFF = drain-source OFF volts

EXAMPLE: What is the power dissipation in a power FET used as the switch element in a discontinuous-mode flyback power supply operating at 30 KHz? The FET operates at a maximum duty cycle of 40%, a peak current of 2.3 amps, a maximum off voltage of 62 volts, a maximum fall time of 100 nanoseconds, and has a maximum on-resistance of 0.65 ohms and no snubber?

PFiDM = clamped inductive load, sawtooth current waveform
(discontinuous mode switching power supply)

= (D/3)*Ids^2*rDS+fS*Ids*VdsOFF*tFI*Ksf
= (0.4/3)*2.3^2*0.65+30E3*2.3*62*100E-9*1
= 0.886 watts

EXAMPLE: What is the power dissipation in the FET from the previous example if the supply is operated at 200 KHz?

PFiDM = (0.4/3)*2.3^2*0.65+200E3*2.3*62*100E-9*1
= 3.310 watts

Table 6.1.1-4
RECTIFIER SWITCHING LOSSES, WATTS

PDrct = rectangular current waveform
= D*Ipk*Vf+0.5*fS*Vr*Ir*tRR*Kd
PDsaw = sawtooth current waveform
= 0.5*D*Ipk*Vf+0.5*fS*Vr*Ir*tRR*Kd

PDped = pedestal+ramp current waveform
= 0.5*D*(Iped+Ipk)*Vf+0.5*fS*Vr*Ir*tRR*Kd

where D = duty cycle
fS = frequency, Hz
Iped = forward pedestal amps
Ipk = peak forward amps
Ir = reverse amps
tRR = reverse recovery time, sec
Vf = forward volts
Vr = reverse volts
Kd = 0.75 if soft recovery or 0.5 if fast recovery diode

EXAMPLE: What is the power dissipation in a fast recovery rectifier used in a discontinuous-mode flyback power supply operating at 30 KHz? The rectifier operates at a secondary maximum duty cycle of 45%, a peak current of 6.5 A, a max forward voltage of 1.1 V, a max reverse voltage of 12 V, a max leakage current of 100 microamps, and a max reverse recovery time of 250 nanoseconds?

PDsaw = sawtooth current waveform
$$= 0.5*D*Ipk*Vf+0.5*fS*Vr*Ir*tRR*Kd$$
$$= 0.5*0.45*6.5*1.1+0.5*30E3*12*100E-6*250E-9*0.5$$
$$= 1.61 \text{ watts}$$

EXAMPLE: What is the power dissipation of a Schottky rectifier used in the power supply from the previous example? The Schottky has a max forward voltage of 0.72 V, a max leakage current of 20 milliamps, and a max apparent reverse recovery time of 75 nanoseconds?

$$PDsaw = 0.5*0.45*6.5*0.72+0.5*30E3*12*20E-3*75E-9*0.5$$
$$= 1.05 \text{ watts}$$

Table 6.1.1-5
SNUBBER RESISTOR POWER, WATTS

PRdamp = C*V^2*f   PRsnub = 0.5*C*V^2*f

where C = snubber capacitor value, farads
      V = voltage of charged capacitor
      f = snubber switching frequency

EXAMPLE: What is the power dissipation of a resistor in a damper circuit with a 0.0022 μF capacitor charged to 60V and discharged through the resistor at a 35 KHz rate?

PRsnub = C*V^2*f
     = 0.0022E-6*60^2*35E3
     = 0.277 watts

EXAMPLE: What is the power dissipation of a resistor in a snubber circuit with a 0.01 μF capacitor charged to 120V and discharged through the resistor at a 50 KHz rate?

PRsnub = 0.5*C*V^2*f
     = 0.5*0.01E-6*120^2*50E3
     = 3.6 watts

6.1.2 Heat Management Basics

After you're sure that you've correctly computed the power dissipation of a component, the next step is to be sure that the component can safely handle the power, which may include the need for a heat sink. The first step is to check the required total *thermal resistance* of the configuration. The total thermal resistance is the sum of all applicable thermal resistances:

Table 6.1.2-1
EXAMPLES OF TOTAL THERMAL RESISTANCE
FOR SEMICONDUCTORS

Rθtot (no heat sink) = Rθja   C/watt
Rθtot (with heat sink) = Rθjc+Rθcs+Rθsa   C/watt

where Rθja = junction-ambient or baseplate thermal resistance, C/watt
Rθjc = junction-case thermal resistance, C/watt
Rθcs = case-sink (insulator) thermal resistance, C/watt
Rθsa = sink-air (heat sink) thermal resistance, C/watt

When the component or its heat sink is mounted to a chassis wall or other structure which serves as an "infinite" heat sink, the mounting surface is generally referred to as the *baseplate*. Ambient or baseplate temperatures are often assumed by definition to be fixed, which implies that the heat generated by the component will have negligible effect on the ambient air or baseplate temperatures. In many instances these assumptions may not be valid.

For a circuit in an enclosure with no forced air (fan) cooling, the air will be substantially confined, allowing the entrapped air temperature to rise as a result of the circuit's power dissipation. Therefore, the *internal* ambient temperature will generally be higher than the enclosure's *external* ambient temperature. For example, a typical well-designed electronics enclosure with "still air" (no fan) will have an internal ambient temperature 5 to 15 degrees C hotter than the external ambient.

THERMAL DESIGN & ANALYSIS TIP # 2
*Be sure to use the **internal** ambient temperature for thermal calculations*

Even when a fan is used, the temperature in different portions of the enclosure may still vary significantly, depending on fan characteristics, venting design, heat sources, etc. In such cases, the temperature at various points in the interior of the enclosure should be measured to confirm that no "dead air" hot spots exist due to design oversights.

Similar considerations apply for baseplate temperatures. If the system design has been done properly, the specified maximum baseplate temperature will already include the effects of expected heat transfer to the baseplate, but *it never hurts* to check.

THERMAL DESIGN & ANALYSIS TIP # 3
Be sure to determine the increase in baseplate temperature which will occur as a result of heat flow from the circuit to the baseplate

---

A special case to watch out for is when a circuit must operate at a high altitude. The lower density atmosphere results in an equivalent higher ambient temperature, since the atmosphere cannot transfer heat as efficiently; i.e., the thermal resistance of the atmosphere increases. In fact, as one goes up to extremely high altitude or space environments, the atmosphere in essence vanishes and the thermal impedance becomes infinite! In such cases, proper thermal management requires that components be mounted on heat sinks of some form.

For high altitude applications, the following tip can be used:

THERMAL DESIGN & ANALYSIS TIP # 4
HOW TO ACCOUNT FOR THE EFFECTS
OF ALTITUDE ON TEMPERATURE
*Add 0.86 degrees to the required upper operating temperature
limit for each additional 1000 feet of altitude above sea level*

EXAMPLE: A circuit's internal ambient temperature specification is 0 to
85 degrees C at sea level. The circuit will be operated at a
maximum altitude of 10,000 feet. What's the maximum design temperature?

$$\text{Tdesign(max)} = \text{Tspec(max)} + 0.86/1000 * 10000$$
$$= 85 + 8.6$$
$$= 93.6 \, C$$

After determining the maximum ambient or baseplate temperature, the
following equations can be used to calculate the expected component temperature:

Table 6.1.2-2
JUNCTION TEMPERATURE FOR TYPICAL CONFIGURATIONS

Temperatures in degrees Centigrade

Temperature of semiconductor junction when part is mounted without heat sink

$$Tj = Ta+R\theta ja*P$$

Temperature of semiconductor junction when part is mounted with heat sink

$$Tj = Ta+(R\theta jc+R\theta cs+R\theta sa)*P$$

Temperature of semiconductor junction when part is mounted on heat sink which contains other power devices

$$Tj = Ta+R\theta sa*Ptot+(R\theta jc+R\theta cs)*P$$

Temperature of transformer or inductor core where cooling path is surrounding air

$$Tc = Ta+P*R\theta ca/Am$$

where Ta = ambient temperature, C
  P = power dissipation of device, watts
  Ptot = total power dissipation of all devices on heat sink, watts
  $R\theta ja$ = junction-ambient thermal resistance, C/watt
  $R\theta jc$ = junction-case thermal resistance, C/watt
  $R\theta cs$ = case-sink (insulator) thermal resistance, C/watt
  $R\theta sa$ = sink-air (heat sink) thermal resistance, C/watt
  $R\theta ca$ = case-air (transformer/inductor) thermal resistance,
      C/W/cm^2 (typical max value = 850 C/W/cm^2)
  Am = surface area, cm^2

EXAMPLE: What is the maximum junction temperature of an LM317A voltage regulator in a TO-220 package which will dissipate a maximum of 1.4 watts, will be operated at a maximum surrounding air temperature of 93.6C, and which has a maximum allowable junction temperature of 125C? Assume that the device will be used without a heat sink.

From the data sheet for the LM317A, the thermal resistance for the TO-220 (T package) without a heat sink (junction-to-ambient) is 50 C/W.

$$Tj = Ta+R\theta ja*P$$
$$= 93.6+50*1.4$$
$$= 163.6\ C$$

Since the junction temperature will exceed the maximum rating of 125C, a heat sink is required.

EXAMPLE: What is the maximum junction temperature of the LM317A voltage regulator from the previous example, assuming that the regulator will be mounted on the heat sink with a thermal resistance of 10 C/W? Assume that the thermal resistance $R\theta cs$ of the mounting insulator for the regulator is a maximum of 0.5 C/W.

From the data sheet for the LM317A, the junction-case thermal resistance for the TO-220 (T package) is 5 C/W.

$$Tj = Ta+(R\theta jc+R\theta cs+R\theta sa)*P$$
$$= 93.6+(5+0.5+10)*1.4$$
$$= 115.3\ C$$

Since the maximum junction temperature is less than 125C, the heat sink is acceptable.

EXAMPLE: What is the maximum junction temperature of the LM317A voltage regulator from the previous example, assuming that the regulator will be mounted on the heat sink along with a power transistor which dissipates a maximum of 2.35 watts?

The Design Analysis Handbook

$$Tj = Ta+R\theta sa*Ptot+(R\theta jc+R\theta cs)*P$$
$$= 93.6+10*(1.4+2.35)+(5+0.5)*1.4$$
$$= 138.8\ C$$

With the extra device mounted on the heat sink, Tj(max) exceeds 125C and therefore the heat sink is inadequate.

If you are determining heat sink requirements, use the following expressions:

Equation 6.1.2-1
REQUIRED THERMAL RESISTANCE OF HEAT SINK

For single device on heat sink:

$$R\theta sa(max) = (Tj(max)-Ta(max))/P(max)-(R\theta jc+R\theta cs)\ C/watt$$

For multiple devices on heat sink:

$$R\theta sa(max) = (Tj(max)-Ta(max)-P(max)*(R\theta jc+R\theta cs))/Ptot(max)\ C/watt$$

where Rθsa(max) = maximum sink-air (heat sink) thermal resistance, C/watt
Tj(max) = maximum allowable junction temperature, C
Ta(max) = maximum ambient or baseplate temperature, C
P(max) = max component power dissipation, watts
Rθjc = junction-case thermal resistance, C/watt
Rθcs = case-sink (insulator) thermal resistance, C/watt

EXAMPLE: What is the maximum allowable heat sink thermal resistance for the LM317A voltage regulator from the previous example if it is mounted alone on a heat sink?

$$R\theta sa(max) = (Tj(max)-Ta(max))/P(max)-(R\theta jc+R\theta cs)\ C/watt$$
$$= (125-93.6)/1.4-(5+0.5)$$
$$= 16.93\ C/W$$

EXAMPLE: What is the maximum allowable thermal resistance of the heat sink for the previous example if the power transistor dissipating 2.35 watts will also be on the heat sink?

Rθsa (max) =
(Tj(max)-Ta(max)-P(max)*(Rθjc+Rθcs))/Ptot(max) C/watt
   = (125-93.6-1.4*(5+0.5))/(1.4+2.35)
   = 6.32 C/W

Note that this satisfies the requirements for the regulator. But what about the power transistor? Assume that the transistor has a maximum allowable junction temperature of 150C, a junction-case thermal resistance of 7 C/W, and an insulator thermal resistance of 0.3 C/W.

Rθsa(max) =
(Tj(max)-Ta(max)-P(max)*(Rθjc+Rθcs))/Ptot(max) C/watt
   = (150-93.6-2.35*(7+0.3))/(1.4+2.35)
   = 10.47 C/W

The heat sink should be selected based on the minimum thermal resistance requirement considering all devices; i.e., 6.32 C/W should be used for the example above.

6.1.3 Thermal Time Constants

When performing thermal calculations, a key consideration is the *thermal time constant* of the component. The thermal time constant Tt is the time that it takes the component to reach 63.2% of its final temperature following the application of a step power pulse. After about three time constants, the temperature has reached about 95% of its final value; after 5 time constants, it's reached more than 99% of its final value. Therefore, if the time constant is *small* with respect to the power pulse width Tp, then the component temperature can actually track the applied instantaneous power:

THERMAL DESIGN & ANALYSIS TIP # 5
For thermal calculations,
Use average power if Tt > 3*Tp
Use instantaneous power if Tt <= 3*Tp
where Tt = thermal time constant
Tp = power pulse period

Thermal time constant information is often shown indirectly by the *transient thermal resistance* rT on the component data sheet. rT is a factor which is multiplied by the steady-state thermal resistance to obtain the effective thermal resistance when using peak powers at a given duty cycle, and is generally provided in the form of a graph of rT versus pulse time for rectangular pulse train waveforms.

EXAMPLE: What is the junction temperature of an MJ13080 power transistor with a junction-case thermal resistance $R\theta jc$ = 1.17 C/W with a rectangular power waveform of 100 watts peak with a pulse width of 1.0 millisecond occurring at a duty cycle of 10%? Assume the transistor is "perfectly" mounted on a heat sink which is maintained at room temperature.

From the MJ13080 data sheet, using the Thermal Response graph, a value of rT = 0.24 is obtained.

$R\theta jc$(transient) = $R\theta jc$*rT
= 1.17*0.24
= 0.281 C/W

Tj = Tc + Ppk * $R\theta jc$(eff)
= 25 + 100 * 0.281
= 53.1 C

6.1.4 Miscellaneous Thermal Management Guidelines

THERMAL DESIGN & ANALYSIS TIP # 6
SOME ADDITIONAL THERMAL MANAGEMENT GUIDELINES

- mount printed wiring boards and heat sinks *vertically* (so maximum surface area has vertical orientation)

- locate power components away from heat-sensitive components

- keep leads as short as possible; e.g., power rectifiers

- run the leads of PWB-mounted power devices to thermal pads (a few square inches per lead of higher weight copper (e.g., 2-ounce) exposed to the air.

- use black heat sinks (significant radiation improvement over plain aluminum)

- guard against deformation and weakening due to thermal cycling by employing materials at mating surfaces with similar temperature coefficients of expansion.

- include circuitry to minimize thermal transients (soft-start, surge limiters, etc.)

- when determining heat sink requirements, also consider whether the heat sink can be touched or whether its heat may damage/degrade nearby components; these factors may require a heat sink larger than one based solely on component junction temperature considerations. Rough rule-of-thumb: if the surface is going to be touched, it should have a max temperature of 60 C (140 F). For momentary inadvertent skin contact, the max temperature should be less than 80 C (176 F).

6.1.5 What About Low Temperature?

Many integrated circuits are rated for "still air" operation, where they are guaranteed to operate properly without a heat sink, provided the ambient air is maintained within the stated limits. For example, a typical IC temperature operating range is Ta = -55 to 125 C. For ICs whose power is not negligible, this implies that the actual junction temperature can be quite a bit higher.

Now, when such an IC is mounted on a printed wiring board, the board itself can act as a heat sink, perhaps resulting in an overall thermal resistance which is significantly *less* than the pessimistic one used by the manufacturer to characterize the IC's performance. This means that the min/max parameter limits provided by the IC vendor may be based on a higher junction temperature range than used in your application. "So what?" you say, since this gives you even more margin at high temperatures. Well, what about *low temperature* performance? By virtue of better heat sinking, your application allows the IC junction to operate in a region which *has not been measured or guaranteed by the vendor.*

Something to think about.

THERMAL DESIGN & ANALYSIS TIP # 7
Don't forget to ensure proper operation at
low temperature *as well as high temperature.*

6.2
AC FULL-WAVE RECTIFIER CIRCUITS

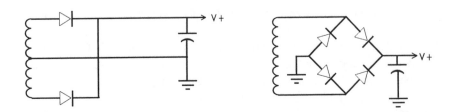

Figure 6.2-1
Full-Wave Rectifier Circuits

This circuit is probably the most used and most under-designed circuit in electronics history. It is not uncommon to find an AC rectifier circuit designed to operate at output currents which will exceed the ratings of its transformer, rectifier diodes, and filter capacitor.

These underdesigned circuits can also be spotted by physically checking for signs of overheating; e.g., burn marks on the board under/around the rectifier package. You can often smell the telltale signs of excessive heat dissipation. But don't check too closely with the power on; an overstressed filter capacitor can explode in your face.

The reason for the underrated components is that a designer mistakenly uses the *average* output current to compute component ratings, rather than the *rms* output current. Since most rectifier circuits are used with a capacitive filter, operating currents will be pulsed, not sinusoidal, resulting in rms values much higher than the average rectified current.

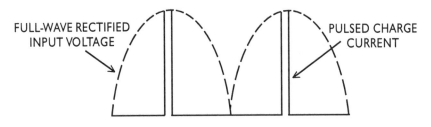

FULL-WAVE RECTIFIED INPUT VOLTAGE   PULSED CHARGE CURRENT

Figure 6.2-2
Charge Current for Full-Wave Rectifier with Capacitive Filter

TIP FOR RECTIFIER CIRCUITS
Use RMS Current Values to Select Components

The problem is compounded because of the tendency of some vendors to rate rectifier diodes for resistive loads as opposed to their more common capacitive load application. An inexperienced designer, upon reviewing a typical rectifier data sheet with its "average rectified DC current" rating, will not receive any warning that the resistive ratings are not appropriate for the application. It would be better if the rectifier's *rms* rating was provided; this would at least require the designer to think about rms rather than average current.

Knowing that you need to design for rms currents is the first step. The second step is calculating the maximum rms current for the various components. For a seemingly simple circuit, this is not a trivial task.

The following equations can be used for full-wave rectified circuits. The equations assume that the source impedance is low compared to the load resistance, which is typically the case.

To determine the rms value of the operating current, the key parameter is the pulse width t of the current pulse. Two expressions are given for t. Use the first if you are designing a new circuit and can specify the desired ripple voltage; use the second if you are analyzing an existing design.

Charge Current Pulse Width

The charge current in the full-wave rectifier circuit with capacitive filter is closely approximated by a rectangular waveform with pulse width t:

Equation 6.2-1
VALUE OF CHARGE TIME t FOR AC
FULL-WAVE RECTIFIER CIRCUIT

$$t = ACOS(1\text{-rRIP})/(2*Pi*f) \text{ seconds}$$

where ACOS = inverse cosine function
rRIP = ripple ratio; e.g., for 15% ripple rRIP = 0.15
f = line frequency, Hz

Base the design on a desired maximum ripple voltage, generally around 15% or so of the minimum rectified voltage. The larger the allowable ripple voltage, the smaller the required filter capacitor, which results in wider current pulses and lower rms currents. (Of course, the ripple can't be *too* large, or you lose the benefits of rectification!)

EXAMPLE: For a desired ripple of 20% and a line frequency of 60 Hz,
t = ACOS(1-rRIP)/(2*Pi*f)
= ACOS(1-0.2)/(2*Pi*60)
= ACOS(0.8)/376.99
= 0.6435/376.99
= 1.71E-3 = 1.71 milliseconds

After solving for t, determine the required value of filter capacitance C:

Equation 6.2-2
VALUE OF FILTER CAPACITANCE FOR
FULL-WAVE RECTIFIER CIRCUIT

C = Io*(0.5/f-t)/dV farads

where Io = DC output current, amps
f = line frequency, Hz
t = charge current pulse width, seconds (EQ 6.2-1)
dV = peak-peak ripple volts

EXAMPLE: For Io = 2 amps, line frequency = 60 Hz, t = 1.71E-3 seconds, and
dV = 20% of a 24V rectified DC output = 4.8V pk-pk,

C = Io*(0.5/f-t)/dV
C = 2*(0.5/60-1.71E-3)/4.8
C = 0.002760 = 2760 microfarads (use nearest standard value)

After defining C, you may wish to calculate the final range of t using the equation in the next section, which will account for all tolerances.

Value of t When Analyzing an Existing AC Full-Wave Rectifier Circuit

If you're analyzing an existing design, the equation for pulse width t will also contain t in the right-hand side of the expression, requiring an iterative solution:

Equation 6.2-3
VALUE OF CHARGE TIME t FOR AC
FULL-WAVE RECTIFIER CIRCUIT
(DESIGN VERIFICATION)

t = ACOS(1-Io*(0.5/f-t)/(Vpk*C))/(2*Pi*f) seconds

where ACOS = inverse cosine function
Io = DC output current, amps
f = line frequency, Hz

C= filter capacitance, farads

$$Vpk = \text{peak rectified line volts}$$

Vpk can be expressed as $1.414*VAC*N-Vrf$ volts

where VAC = AC line volts
N = secondary-to-primary turns ratio
Vrf = forward voltage drop of rectifier(s)

EXAMPLE: For Io = 2 amps (max), f = 50 to 60 Hz, C = 2700 μF +/- 20%, VAC = 110 to 130 VAC, N = 0.15, and Vrf = 1.5 to 2.5 volts,

$$t = ACOS(1-Io*(0.5/f-t)/((1.414*VAC*N-Vrf)*C))/(2*Pi*f)$$
seconds

Solving the above iteratively,

t min = 1.53 milliseconds (f=60, VAC=130, Vrf=1.5, C=3240E-6)

t max = 2.65 milliseconds (f=50, VAC=110, Vrf=2.5, C=2160E-6)

Component Ratings of AC Full-Wave Rectifier Circuit

After calculating the value of t, use the following equations to determine the component ratings for the circuit. The equations are dependent upon the duty cycle of the charge current pulse, where the duty cycle is given by $2*t*f$. The worst case maximum rms current is given by the minimum duty cycle.

Equation 6.2-4
TRANSFORMER SECONDARY RMS CURRENT

Ixfmr = Io/SQR(2*t*f) amps rms

where Io = average DC output amps
SQR = square root function
t = charge current pulse width, seconds (EQ 6.2-1)
f = line frequency, Hz

EXAMPLE: For the values from the prior example,
Ixfmr = Io/SQR(2*t*f)
= 2/SQR(2*1.53E-3*60)
= 4.67 amps rms

Note that the rms current is more than double the average output current.

Equation 6.2-5
RECTIFIER RMS CURRENT

Irect = 0.5*Io/SQR(2*t*f) amps rms

where Io = average DC output amps
SQR = square root function
t = charge current pulse width, seconds (EQ 6.2-1)
f = line frequency, Hz

For the full-wave rectifier configuration, each diode is operated at a 50% duty cycle, hence the rms amps are 1/2 that of the transformer.

EXAMPLE: For the values from the prior example,

Irect = .5*Ixfmr = 2.33 amps rms

The Design Analysis Handbook

Note that since rectifier ratings are often only given in terms of average current, you will need to obtain the rms rating from the vendor, which may be on the order of 1/2 or less than the average rating. Also, check the data sheet curves; some vendors provide derating graphs for capacitive load applications.

You will also want to check the rectifier's *surge rating*; this the the maximum allowable peak current during the initial charge of the filter capacitor:

Equation 6.2-6
RECTIFIER SURGE CURRENT

Isurge = Vac*1.414/Zckt

where Vac = rms AC input volts
Zckt = impedance of charge circuit

Input filters and transformers generally have sufficient leakage inductance and series resistance to keep Zckt high enough to avoid problems. For exceptions to this (e.g., off-line rectifiers without input filters) Zckt is limited only by the output filter capacitor's ESR and the rectifier's internal dynamic impedance, plus very small line and wiring impedances. In such cases a hefty surge rating may be required, or alternately Zckt can be augmented with a thermistor or other inrush current-limiting methods.

Small resistors are sometimes used in series with the rectifier to limit inrush current, but the power in a series resistor can be surprisingly high, even for small-ohm values (see example below).

Equation 6.2-7
SURGE RESISTOR POWER

PR = (Io/SQR(2*t*f))^2*R watts

where Io = average DC output amps
SQR = square root function
t = charge current pulse width, seconds (EQ 6.2-1)

f = line frequency, Hz
R = value of any resistor used in series with rectifier

EXAMPLE: For the values from the prior example, and assuming a 0.5
ohm surge resistor,
PR = (Io/SQR(2*t*f))^2*R
= (2/SQR(2*1.53E-3*60))^2*0.5
= 10.89 watts

Equation 6.2-8
CAPACITOR RMS CURRENT

Icap (charge current) = ICchg = Io*SQR(1/(2*t*f)-1) amps rms

where Io = average DC output amps
SQR = square root function
t = charge current pulse width, seconds (EQ 6.2-1)
f = line frequency, Hz

EXAMPLE: For the values from the prior example,

Icap = Io*SQR(1/(2*t*f)-1)
= 2*SQR(1/(2*1.53E-3*60)-1)
= 4.22 amps rms

For the capacitor, you may also need to add the discharge ripple current
ICdis if the load draws pulsed load current; e.g., switching power supply.
If such is the case, then the total ripple current will be

ICrms = SQR(ICchg^2+ICdis^2)

6.3
POWER SUPPLIES

Power supply design is one of the most demanding engineering challenges. Switchmode power supplies, in particular, are exceptionally difficult. Switchers embody high-level high-speed power switches, magnetics design, thermal management, low-level signal processing in a noisy environment, one or multiple control loops and attendant response, stability issues, and more. Plus (as every experienced switcher designer knows), it has to deliver more power, have higher efficiency, and be smaller than the specification called for last month. (...and by the way, have the design completed by Thursday.)

There are lots of good books available for power supply design (see references section), so this section will focus primarily on a few of the topics which tend to be more subtle and have the potential for creating serious problems.

6.3.1 Specifications

One of the problems right off the bat with power supplies is that the specifications often needlessly tie the hands of the designer. The person writing the spec should not feel compelled to define a separate requirement for every operational variable (line, load, noise, etc.), but should specify an *allowable total tolerance for each output*. After all, in most cases the user simply wants to be sure that a given output voltage will be within a specified tolerance for all combinations of operational variables. Why not save some time and money and give the designer the freedom of a total error budget?

WHY NOT HAVE A KEY POWER SUPPLY SPEC KIND OF LIKE THIS?
*Each output voltage shall be within its defined total
allowable tolerance for the combined effects of the full
range of component tolerances, temperature, line voltage,
load current, and peak noise plus ripple, for the required
operating lifetime of the power supply.*

6.3.2 Stability with an Input Filter

Since a switchmode power supply has a negative input impedance, it's possible for an otherwise stable supply to oscillate when connected to an input (EMI) filter, depending on supply and filter characteristics. That's one good reason for having the supply and the filter designed together, rather than trying to add an input filter later.

POWER SUPPLY STABILITY WITH AN INPUT FILTER

Make sure that a switchmode power supply and its EMI input filter meet the following constraints:

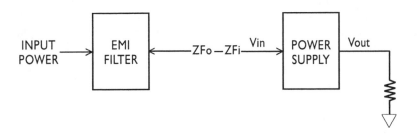

1. $ZFo \ll -R/M^2$

2. $ZFo \ll ZPi/M^2$

where ZFo = magnitude of input filter's output impedance, ohms
R = output load ohms of power supply
M = Vout/Vin = output voltage / input voltage of power supply
ZPi = magnitude of reflected input impedance of power supply (including effect of R), ohms

---

The above constraints must be satisfied for all frequencies; this is generally accomplished by having a large capacitor at the output of the input filter. Also, it is assumed that the power supply is quite stable *without* the input filter present.

6.3.3 Duty Cycle Limit

A switchmode power supply is defined by its conduction mode, which for the common duty cycle controlled varieties is either *continuous* or *discontinuous*. For the same physical hardware, performance can change dramatically depending upon which mode the supply is in. The mode is determined by the supply's design and its duty cycle. Therefore, if operating conditions change enough to push the duty cycle over the *conduction boundary*, then performance may abruptly change, resulting in unstable operation. Power supply designers are well aware of these factors, but sometimes get zapped by transient conditions, which can bump the duty cycle up beyond normal design limits, and kick the supply into an unstable mode.

Duty cycle variations which exceed the normal operating range can also result in transformer saturation, which is highly undesirable for most converter types. If the transformer saturates, the power devices used to drive the transformer will see a short circuit, resulting in loss of regulation and even (depending on power supply topology) component failure. The design should ensure that the core will not saturate under worst case conditions, including startup, shutdown, step load, and other transient cases. Core saturation can also be caused by "flux walking" which results from driving the transformer with an asymmetrical waveform. A hardware duty cycle limit will also help prevent core saturation.

POWER SUPPLY DUTY CYCLE LIMIT

Use a hardware duty cycle limit to:

- prevent unintended conduction-mode changes and related loss of control or oscillations.

- help prevent core saturation and related malfunctions and component failures.

For example, a flyback power supply has a well-defined stable operation as long as it is in the discontinuous mode. As the duty cycle increases, it approaches the continuous mode boundary. If it enters the continuous mode, the loop performance becomes much more complex, with compensation depending upon operating conditions (the "wandering pole" phenomenon), and instabilities are likely (unless designed for continuous mode...a tough challenge). Therefore, it's wise to incorporate a duty cycle limit feature in the design, which guarantees that the supply will never enter an undesired mode, even in the presence of transients.

With a duty cycle limit, the power supply should of course be designed such that the limit is never reached under normal operation, otherwise the output voltage will drop out of regulation. For transient conditions, if the supply hits the duty cycle limit the output may droop temporarily, but control will not be lost.

6.3.4 Layout

Because a switcher generates high-amplitude high-speed waveforms near low-level control circuitry, proper layout is a necessity to prevent oscillations due to parasitic coupling, to prevent output voltage inaccuracies due to noise pickup, and to minimize EMI. Also see Section 6.5, "Grounding and Layout," and Section 6.6, "EMI / Noise Control."

POWER SUPPLY LAYOUT TIPS

- twist together any cabled power leads which supply raw power to the input of the power supply.

- keep each input power loop as small as possible (the input loop is the path from the input filter capacitor through the transformer primary through the power switch(es) to the input filter ground).

- keep each output power loop as small as possible (an output loop is the path from the transformer secondary through the rectifier through the filter capacitor to the secondary common).

- any low-level high impedance circuitry in the vicinity of a switching power supply (including the supply's own control circuit) should be laid out in a minimum area and located away from the supply's high-current paths. A high impedance path of adequate dimensions can act as an antenna and readily detect the power supply's high frequency switching transients. If a diode or other junction is in the loop, the loop can act as a noise pump, eventually creating large error voltages from a repetition of small noise impulses.

• intolerable EMI noise has been observed to be generated by chassis-mounted switching transistors. The cure is to mount the switching devices on an electrically-isolated heat sink, or to mount them using low-capacitance insulators; i.e., the goal is to minimize the capacitive coupling between the transistor cases and the chassis.

6.3.5 Thermal Management

It's really disappointing to complete a great paper design, and then have the prototype exhibit exploding capacitors, flaming resistors, or lid-popping power devices, particularly when your boss is watching. While we've witnessed dramatic melt-downs such as these, in general most thermal problems we've detected have been because the components used in a switcher were not adequate for their application, resulting in long-term reliability problems; i.e., a slow burn instead of an explosion. These defects are usually a result of the designer not correctly computing the power that a component must dissipate. Discontinuous mode topologies (e.g., the common flyback power supply), are especially prone to problems because of the high rms currents inherent in their pulsed current waveforms.

Trouble-spots include input/output filter capacitors, output rectifiers, and snubbers. Schottky diodes can be troublesome because their high reverse leakage currents may lead to thermal runaway.

To keep things cool, calculate the correct power for power supply waveforms by using the appropriate equations from Section 6.1, "Power and Thermal." Also, a handy tip is provided below:

TIP FOR PREVENTING CHAIN REACTION DAMAGE
*Use flameproof (fusible) resistors in series with key power
circuits to limit the damage from overload failures.*

6.3.6 The Unloaded Flyback and Other Hazards

DO'S and DON'TS of POWER SUPPLY DESIGN

- Never allow a flyback power supply to operate with no load (you'll
  blow up the output capacitors). If it is possible for the power supply
  output to be disconnected from its load while being operated or tested,
  then include a resistor or zener directly across the output filter to prevent a no-load condition when disconnected.

- A shorted pass transistor or out-of-control regulator can destroy all
  components on the power bus...not a pretty sight. Prevent such catas-
  trophes by using a clamping zener or other overvoltage protection
  device at each output. Such protection is particularly valuable at test
  station power supply outputs.

- PWM ICs used to control switchmode power supplies may sometimes fail to lock to their synchronization signals. The reason is that
  the center frequency of some PWM ICs has a broad initial value,
  even if precision frequency-setting components are used. To ensure
  proper centering, you may have to include a frequency-trimming resistor.

- Lack of power supply synchronization in a video display sometimes
  exhibits itself as low-level but annoying video jitter ("swimming pixels"), possibly accompanied by a soft hissing sound.

- Be sure to have enough filter capacitance for line-rectified sources
  feeding three-terminal regulators to avoid dropout problems. If the
  input ripple is too large, the negative peaks may cause periodic dips
  in the regulator output. Always use the most negative peak input

voltage when calculating dropout requirements, not the minimum DC (average) input voltage.

- When checking the breakdown capability of a three-terminal regulator, the maximum input-output operating volts may not tell the whole story. The worst case voltage occurs when the regulator is off (output filter capacitor discharged) and the input voltage is suddenly applied. In such cases, the regulator must be rated for the maximum input voltage, or external protection (e.g., zener) must be added.

The Case of the Unsynchronized Power Supplies

Vexing intermittent EMI can occur when multiple switchmode power supplies are operated at closely spaced but unsynchronized frequencies; the modulation products of the frequencies can create upper and lower noise sidebands. While the higher frequencies are relatively easy to filter or shield, the lower frequencies are not, and can create serious interference problems that "come and go" in a manner sure to challenge the patience of the most diligent troubleshooter:

AVOIDING THOSE BEAT FREQUENCY BLUES
*Synchronize switchmode power sources to each other
and to other switching power circuits.
Hint: Synchronize the circuits in frequency (or to harmonics)
but not in phase; a fixed phase offset is desirable to
prevent simultaneous large current pulsing.*

6.3.7 Leakage Inductance

Has this ever happened to you? Your power supply has been working great, but the latest batch off the production line all fail. You check with purchasing and find that the power transformer has been purchased from a different vendor to save money. Ah-ha! So you take a sample of the new transformer back to receiving inspection and test it, but it meets all the specs. Hmmm.

We'd be willing to bet that the spec doesn't contain either (a) a maximum limit for *leakage inductance,* or (b) physical winding instructions which control the same.

Leakage inductance is the parasitic inductance which results from imperfect coupling between transformer windings. For switchmode power supplies, leakage inductance is critical...too much, and you'll never get the supply to work properly. As a rule-of-thumb, leakage inductance should be a small fraction of the associated primary of secondary inductance; e.g., less than 3%.

TIP FOR ENSURING REPEATABLE
TRANSFORMER PERFORMANCE
Be sure to specify the maximum allowable leakage inductance.

6.3.8 Voltage Fed Converters

Numerous failures have been observed in voltage-fed power converters. A voltage-fed converter receives its power from a low impedance source, and the converter itself contains no inherent current-limiting impedances. Without such current limiting, transients or faults can result in extremely high currents which damage the converter. Many push-pull designs have a start-up behavior which can temporarily allow simultaneous conduction of their power switches, creating very high transient currents. Since these supplies tend to fail when they are turned on (like light bulbs do), their failures sometimes seem mystifying.

Various design approaches can be employed to prevent push-pull failures, including improved startup circuits and current-sense circuits which shut off the power switches in the event of overcurrent detection. However, the simplest solution is to simply avoid the voltage-fed topology and use a current-fed topology instead.

A current-fed converter employs an inductor in the supply line which acts as a short-term current source, thereby hardening the power supply against the effects of simultaneous conduction. The inductance may an integral part of the converter topology (e.g., push-pull flyback).

A TIP ON THE USE OF VOLTAGE-FED CONVERTERS
Avoid this topology if you can.

6.4
DIGITAL APPLICATIONS

Digital design problems are commonly identified by Worst Case Analysis despite the plethora of design validation tools available to digital designers (logic validation, timing analysis, waveform control, etc.). Paradoxically, part of this is probably due to the fact that simulation tools can in some instances actually encourage errors because such tools tend to discourage critical thinking; i.e., plugging in a bunch of numbers and seeing what pops out is not really engineering.

Digital problems also occur because in many cases digital circuits actually contain a lot of analog considerations with which many digital designers are unfamiliar. Such analog effects are much more noticeable in the interior of a digital IC, where waveforms in many cases are barely digital at all, and the concepts of slew rate, bandwidth, offsets, etc. become of paramount importance.. Even outside the chip, the digital systems designer can still be stung by an unfamiliarity with the waveform distortion caused by transmission line effects, by noise due to crosstalk, by grounding and layout problems, and other analog concerns.

This section is going to address a couple of basic and important digital design areas which seem to snare more than a few unwary designers. The first is the timing margins analysis; the second is transmission line effects. Some miscellaneous tips are provided at the end of the section.

6.4.1 Timing Margins Analysis

6.4.1.1 Key Timing Margin Parameters

Critical timing paths are constrained by a SETUP margin mSU and a HOLD margin mHD. As will be seen in the equations to follow, the setup margin is a function of frequency, but the hold margin is not. The latter fact may allow major design bugs to be carried in the designs of those who think that a timing margins analysis is unnecessary for "low speed" applications.

TIMING MARGINS ANALYSIS TIP
Hold Margin violations can occur at any operating frequency; i.e., don't dismiss the need for a timing margins analysis for "low-speed" designs.

Setup and hold margins are calculated at the inputs of clocked integrated circuits which have setup and hold timing constraints.

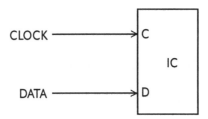

Figure 6.4.1.1-1
Clocked Device with Setup and Hold Constraints

The incoming data must arrive and be stable for a SETUP time tS (specified by the IC's data sheet) prior to the arrival of the clock edge (rising or falling depending on device type). The data must remain present and stable for a HOLD time tH (specified by the IC's data sheet), following the occurrence of the clock edge. The actual time which exceeds the setup requirement is the setup margin mSU. Likewise, the actual time which exceeds the hold timing requirement is the hold margin mHD. For an IC which is clocked by a rising edge, the relationships are shown below:

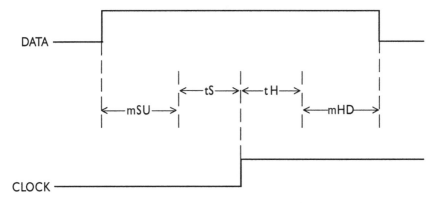

Figure 6.4.1.1-2
Setup tS, Hold tH, Setup Margin mSU, and Hold
Margin mHD Times for Device Clocked by Rising Edge

For correct operation, mSU and mHD must both be positive values. Their equations are:

Equation 6.4.1.1-1
TIMING MARGINS

SETUP MARGIN = mSU = CLK - DAT + PER - OFF - tS

HOLD MARGIN = mHD = DAT - CLK + OFF - tH

where CLK = sum of delay times in clock path
DAT = sum of delay times in data path
PER = clock period
tS = required setup time
tH = required hold time
OFF = offset time of originating data relative to clock

Note that the frequency of operation (1/PER) has an effect on the setup margin, but not on the hold margin.

EXAMPLE: Determine the timing margins for the circuit below, which is operating at a frequency of 100 MHz.

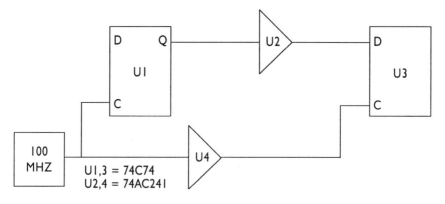

Figure 6.4.1.1-3
Timing Margins Analysis Sample Circuit

The following data sheet values are known:

Device	Parameter (typical)	
U1	Delay from C to Q	= 6.0 nanoseconds (ns)
U2	Delay from in to out	= 4.8 ns
U4	Delay from in to out	= 4.8 ns
U3	Required setup time	= 1.0 ns
U3	Required hold time	= -1.5 ns

Therefore,

$$\begin{aligned}
CLK &= \text{delay of U4} = 4.8 \text{ ns} \\
DAT &= \text{delay of U1} + U2 = 6.0 + 4.8 = 10.8 \text{ ns} \\
PER &= 1/FREQ = 1/100E6 = 10 \text{ nanoseconds} \\
OFF &= 0 \text{ since the same clock edge is used to create the origi-} \\
&\quad \text{nating data waveform as is used to clock the IC (no in-} \\
&\quad \text{versions)} \\
tS &= 1.0 \text{ ns} \\
tH &= -1.5 \text{ ns}
\end{aligned}$$

$$\begin{aligned}
\text{Setup Margin mSU} &= CLK - DAT + PER - OFF - tS \\
&= 4.8-10.8+10-0-1 \\
&= 3.0 \text{ ns}
\end{aligned}$$

$$\begin{aligned}
\text{Hold Margin mHD} &= DAT - CLK + OFF - tH \\
&= 10.8-4.8+0-(-1.5) \\
&= 7.5 \text{ ns}
\end{aligned}$$

Since both timing margins are positive, the timing restrictions for U3 are satisfied.

In the preceding example, typical values were used for all parameters. Note that the typical value for the hold time tH is a *negative* number. Such negative numbers are completely valid and are no cause for alarm; they merely represent the difference between delay paths internal to the integrated circuit.

The Design Analysis Handbook

In practice, the CLK and DAT parameters have a range of min/max values. In such cases, an inspection of the equations for mSU and mHD tells which values to use for worst case results:

Equation 6.4.1.1-2
WORST CASE TIMING MARGINS

suffix m = minimum, x = maximum

SETUP MARGIN = mSUm = CLKm - DATx + PERm - OFFx - tSx

HOLD MARGIN = mHDm = DATm - CLKx + OFFm - tHx

EXAMPLE: Determine the timing margins for the prior example, using the range of min/max values as given below:

Device	Parameter (Min/Max)
U1	Delay from C to Q = 2.5 to 10.5 ns
U2	Delay from in to out = 1.0 to 7.5 ns
U4	Delay from in to out = 1.0 to 7.5 ns
U3	Required setup time = -1.0 to 3.0 ns
U3	Required hold time = -3.5 to 0.5 ns

Therefore,

CLK = delay of U4 = 1.0 to 7.5 ns
DAT = delay of U1 + U2 = 2.5 + 1.0 to 10.5 + 7.5 = 3.5 to 18 ns

Setup Margin mSUm = CLKm - DATx + PERm - OFFx - tSx
= 1.0-18+10-0-3
= -10.0 ns

Hold Margin mHDm = DATm - CLKx + OFFm - tHx
= 3.5-7.5+0-0.5
= -4.5 ns

Using the full range of possible delay values, neither timing restriction for U3 is satisfied; i.e., for those times when tolerances combine to give the worst case combination of delay values, circuit malfunction will occur (U3 will not properly accept incoming data).

If the worst case results are negative as in the example above (timing margin violation), then a probability analysis can be applied to the results to estimate the percent of time that the violation(s) will occur. A sample calculation is provided below, which shows that the setup violation will occur an estimated 7.1% of the time, while the hold violation will only occur 0.007% of the time. The distribution of delays for U1,2,4 were all modeled as approximately normal, truncated at the +/- 3-sigma points.

SAMPLE TIMING MARGINS RESULTS
INCLUDING PROBABILITIES

RECORD #1: mSU = Status: Alert
ALERT probability = 7.0959 %

Setup margin, ns	Minimum	Average	Maximum
Specification............	0		50
Actual......................	-10	2.5	15

```
              |     x                              |
              |     x                              |
              L    xxx                             L
              I   xxxxx                            I
              M  xxxxxxx                           M
               m xxxxxxxxx                         x
             xx | xxxxxxxxxxxx                     |
xxxxxxxxxxxxxxxxxxxx | xxxxxxxxxxxxxxxxxxxxxxxxxxxxxxxxxxx  |
^_____^____
M                                                   M
I                                                   A
N                                                   X
```

Distribution weight = 8.75, centering = .50
Standard Deviation = 1.732167

──────────SOLUTIONS──────────

$$mSU = CLK - DAT + PER - OFF - tS$$
$$= (tU4)-(tU1+tU2)+(10)-(0)-tSSS$$
$$MIN = (1)-(18)+(10)-(0)-(3) = -10$$
$$MAX = (7.5)-(3.5)+(10)-(0)-(-1) = 15$$

RECORD # 2: mHD = = = = = = = = = = = = = = = STATUS: CAUTION
CAUTION probability = 6.555669E-03 %

Hold margin, ns	Minimum	Average	Maximum
Specification............	0		50
Actual.....................	-4.5	8	20.5

```
        |                 x                        |
        |                 x                        |
        L               xxx                        L
        I              xxxxx                       I
        M             xxxxxxx                       M
        m            xxxxxxxx                       x
        |          xxxxxxxxxxxxxx                   |
xxxxxxxxx|xxxxxxxxxxxxxxxxxxxxxxxxxxxxxxxxxxxxxxxxxxxxx  |
_____^_____^____
  M                                               M
  I                                               A
  N                                               X
```

Distribution weight = 8.75, centering = .50
Standard Deviation = 1.732167

──────────SOLUTIONS──────────

$$mHD = DAT - CLK + OFF - tH$$
$$= (tU1+tU2)-(tU4)+(0)-tHHH$$
$$MIN = (3.5)-(7.5)+(0)-(.5) = -4.5$$
$$MAX = (18)-(1)+(0)-(-3.5) = 20.5$$

In the prior example, delay times were considered to be strictly a function of the integrated circuit internal delay times. Actually, the delay time for a given path is a function of several parameters:

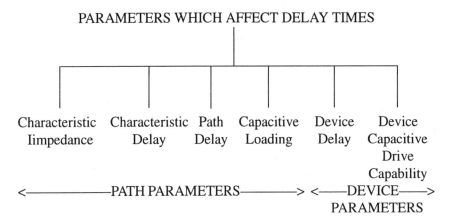

PARAMETERS WHICH AFFECT DELAY TIMES

The two basic delay parameter categories are PATH and DEVICE. We have already discussed *device delay*. A second device parameter is *capacitive drive capability*, which is the IC's capability of driving a path with capacitive loading. When an integrated circuit tries to drive a capacitive load, the load capacitance will slow the rise/fall times at the output of the IC, creating an additional effective delay. This will be discussed in more detail later.

Physical path characteristics also have a significant impact on delay times. The path itself has an intrinsic propagation delay time/length (*characteristic delay*) which is a function of the characteristic impedance of the path and is modified by the capacitive loading of ICs connected to the path. Also, the path has an inherent distributed capacitance which adds to the total IC load capacitance.

6.4.1.2 Characteristic Impedance

The first thing one needs to know in determining the various delay effects is the *characteristic impedance* of the paths, defined below:

The Design Analysis Handbook

Equation 6.4.1.2-1
PATH CHARACTERISTIC IMPEDANCE

$$Zo = SQR(LO/CO) \text{ ohms}$$

where   SQR = square root function
     LO = path inductance/length
     CO = path capacitance/length

Most digital designs are implemented on a printed wiring board (PWB). Generally, the easiest way to obtain Zo is to simply ask the PWB vendor. Since Zo will vary as a function of the path geometry and placement relative to a power or ground plane, ask for Zo values for the range of path types/placements used in the design.

If for some reason you can't obtain Zo from the vendor, the following approximations can be employed:

Equation 6.4.1.2-2
CHARACTERISTIC IMPEDANCE OF
PRINTED WIRING BOARD PATH

Zo (microstrip) = (87/SQR(DCON+1.41))*LN(5.98*H/(0.8*W+T)) ohms

Zo (stripline) = (60/SQR(DCON))*LN(4*B/(0.67*Pi*W*(0.8+T/W))) ohms

where   LN = natural log function
    DCON = dielectric constant of printed wiring board
     H = layer spacing (distance between path and ground plane)
     T = thickness of path
     B = distance between planes = 2*H+T

"Microstrip" refers to the case where the printed wiring path is routed close to a single ground plane; "stripline" refers to the case where the printed wiring path is sandwiched between two planes.

EXAMPLE: What is the characteristic impedance of a microstrip path on a PWB with a dielectric constant of 4.5, H = 20 mils (0.02 inches), W = 10 mils, and T = 1.5 mils?

$$Zo = (87/SQR(4.5 + 1.41))*LN(5.98*.02/(.8*.01 + .0015))$$
$$= 90.6 \text{ ohms}$$

EXAMPLE: Repeat the above example for a stripline path.

$$Zo = (60/SQR(4.5))*LN(4*.0415/(.67*Pi*.01*(.8 + .0015/.01)))$$
$$= 59.9 \text{ ohms}$$

Equation 6.4.1.2-3
CHARACTERISTIC INDUCTANCE OF
PATH OVER GROUND PLANE

$$LO = K*(LN(5.98*H/(0.8*W+T)))$$

where   K = 1.97E-7   for LO = henries/meter
          = 5.0   for LO = nanohenries/inch
       LN = natural log function
        W = width of path
        H = layer spacing (distance between path and ground plane)
        T = thickness of path

EXAMPLE: Find the characteristic inductance of the microstrip path from the previous example:

width W   = 10 mils
layer spacing H = 20 mils
path thickness T = 1.5 mils

$$LO = K*(LN(5.98*H/(0.8*W+T)))$$
$$= 5*(LN(5.98*0.020/(0.8*0.01+.0015)))$$
$$= 12.66 \text{ nanohenries/inch}$$

Equation 6.4.1.2-4
CHARACTERISTIC CAPACITANCE WHEN Zo IS KNOWN

$$CO = LO/Zo^2 \text{ capacitance/length}$$

where   LO = characteristic inductance/length
                  Zo = characteristic impedance, ohms

EXAMPLE: Find the characteristic capacitance of the microstrip path from the previous example:

$$
\begin{aligned}
CO &= LO/Zo^2 \\
&= 12.7E\text{-}9/90.6^2 \\
&= 1.54 \text{ picofarads/inch}
\end{aligned}
$$

6.4.1.3 Characteristic Delay

A signal travels down a path at a rate called the *characteristic delay*, given in time/length. This delay is defined for an unloaded path, therefore the characteristic delay needs to be adjusted in accordance with the capacitance added by connecting one or more ICs to the path.

Formulas for the characteristic delay of common printed wiring board configurations are given below:

Equation 6.4.1.3-1
CHARACTERISTIC PATH DELAY

TPD (microstrip) = 0.085*SQR(0.475*DCON+0.67) nanoseconds/inch

TPD (stripline)   = 0.085*SQR(DCON) nanoseconds/inch

TPD (general)   = SQR(LO*CO) = CO*Zo = LO/Zo

where   SQR = square root function
               DCON = dielectric constant of printed wiring
                      board, (typical value = 4.5)

LO = path inductance/length (EQ 6.4.1.2-3)
CO = path capacitance/length (EQ 6.4.1.2-4)

EXAMPLE: What is the characteristic delay of the microstrip path from the previous example?

TPD = 0.085*SQR(0.475*DCON+0.67)
= 0.085*SQR(0.475*4.5+0.67)
= 0.142 ns/inch

EXAMPLE: Repeat the above example for a stripline path.

TPD = 0.085*SQR(DCON)
= 0.085*SQR(4.5)
= 0.180 ns/inch

The characteristic path delay above is for an "unloaded" line; i.e., a path which does not yet have any receiving devices connected. In practice of course, one to several receiving devices will be connected to a given path. Their combined input capacitance changes the effective path capacitance, as defined by a capacitance load factor CLF:

Equation 6.4.1.3-2
CAPACITANCE LOAD FACTOR

CLF = SQR(1+(Ccomp/PL)/CO)

where   Ccomp = total capacitance of components connected to the path
PL = path length
CO = path capacitance/length (EQ 6.4.1.2-4)

The load factor is used to compute the adjusted path delay:

Equation 6.4.1.3-3
ADJUSTED PATH DELAY

$$TPDA = TPD*CLF$$

where   TPD = characteristic path delay, time/length
                    CLF = capacitance load factor (EQ 6.4.1.3-2)

EXAMPLE: Find the adjusted path delay of a 14-inch microstrip line with total component loading of 13.5 picofarads, using the values for LO and CO calculated earlier:

Ccomp = 13.5 picofarads
TPDA = TPD*CLF
CLF   = SQR(1+(Ccomp/PL)/CO) = SQR(1+ 13.5/14)/1.54)
         = 1.27
TPDA = 0.142*1.27
         = 0.182 ns/inch

6.4.1.4 Path Length

The path's characteristic delay is defined in terms of delay/length. Therefore, the total intrinsic path delay is its characteristic delay multiplied by its length:

Equation 6.4.1.4-1
PATH INTRINSIC DELAY

$$PDEL = TPDA * PL$$

where   TPDA = adjusted characteristic path delay/length (EQ 6.4.1.3-3)
               PL = path length

EXAMPLE: Find the intrinsic path delay of the 14-inch microstrip line from the previous example.

PDEL = TPDA * PL
        = 0.182 ns/inch * 14 inches
        = 2.54 ns

Likewise, the path's characteristic capacitance is defined in terms of capacitance/length, and the total capacitance contributed by the path itself is its characteristic capacitance times its length:

Equation 6.4.1.4-2
PATH INTRINSIC CAPACITANCE

PCAP = CO * PL

where   CO = path capacitance/length (EQ 6.4.1.2-4)
           PL = path length

EXAMPLE: Find the intrinsic capacitance for the 14-inch microstrip line from the previous example:

PCAP = CO * PL
        = 1.54 pf/inch * 14 inches
        = 21.6 pf

6.4.1.5 Capacitive Loading

The loading of a path is the sum of the input capacitances of receiving ICs, the input/output capacitances of bidirectional (tri-state) ICs, the capacitance of any other components connected to the path, plus the intrinsic capacitance of the path itself.

Integrated circuit input capacitances can range from around 4 to 15 picofards per pin. Pins connected to bidirectional devices generally have higher capacitances; 15 to 20 picofarads or so. These capacitance values can be found in the IC data manuals. Intrinsic path capacitances are typically in the range of a few picofarads per inch, so path capacitances will run in the range of a few to tens of picofards, depending on path length.

Equation 6.4.1.5-1
TOTAL LOAD CAPACITANCE

$$Cload = Ccomp + PCAP$$

where   Ccomp = total capacitance of components connected to path
            PCAP = intrinsic path capacitance/length (EQ 6.4.1.4-2)

EXAMPLE: Find the total capacitance for the 14-inch microstrip line from the previous example:

$$Cload = Ccomp + PCAP$$
$$= 13.5 + 21.6$$
$$= 35.1 \text{ pf}$$

6.4.1.6   Device Delay

Device delay is the time required for a signal to propagate internally through an IC. Device delays will generally vary as a function of initial tolerance, power supply, temperature, and capacitive loading. The data sheet usually specifies a min/max range of values for the first three parameters lumped together in the data sheet, for a specified value of load capacitance.

EXAMPLE: What is the clock-to-output device delay for a 74AC74 flip-flop IC?

From the Motorola data sheet for the MC74AC74, the clock to output delay can vary from 2.5 to 10.5 ns for an operating temperature range of -40 to +85 C, a supply voltage of +5V +/-10%, and a load capacitance of 50 pf.

EXAMPLE: What is the clock-to-output device delay for a 74HC74A flip-flop IC?

From the Motorola data sheet for the MC74HC74A, the clock to output delay is specified as a *maximum* of 30 ns for an operating temperature range of -55 to +125 C, a supply voltage of +4.5V to +6.0V, and a load capacitance of 50 pf.

Note that a *minimum* delay spec is not provided, which indicates that the creators of this particular Motorola manual do not understand worst case timing requirements, where minimum delays are just as important as maximum delays. Since no guidance is provided in the data sheet (not even a typical delay value), the designer is left with having to guess at a reasonable minimum delay value.

In such cases, the vendor's applications engineering department should be contacted to obtain the required data. For preliminary calculations, the following guidelines can be employed:

Estimating the Minimum if Maximum and Typical are Known:

Min (est) = Typ - (Max - Typ)
            = 2 x Typ - Max

Estimating the Minimum if Only the Maximum is Known:

Assuming a parameter tolerance of approximately 30% for ICs, the minimum can be estimated as 30% less than the typical, while the maximum is 30% higher than the typical, or

Min (est) = 0.7 x typical
            = 0.7 x maximum/1.3
            = 0.7/1.3 x maximum
            = 0.54 x Max

The Design Analysis Handbook

6.4.1.7 Device Capacitive Drive Capability

The effective delay through an IC is greatly affected by capacitive load-ing of the output. This loading is the sum of the capacitances of all of the receiving device inputs, plus capacitance contributed by the physical path which connects the devices. Therefore, the data sheet value for delay through a device needs to be modified by loading effects. A formula for defining the total device delay is:

Equation 6.4.1.7-1
TOTAL DEVICE DELAY

$$DDEL = DEL + DFACT * (Cload - Ctest)$$

where   DEL   = device delay from data sheet
DFACT   = device capacitive drive capability, delay time
per load capacitance
Cload   = total load capacitance (component+path)
Ctest   = test capacitance for DEL

To obtain DFACT, first check the component data sheet for a graph of delay time versus capacitive loading. If no graph is provided, check the tabular data; sometimes delays are specified for min/max capacitive loads, from which you can estimate DFACT:

Equation 6.4.1.7-2
ESTIMATING DFACT FROM DATA SHEET MIN/MAX VALUES

suffix m = minimum, x = maximum

$$DFACT = (DELx - DELm) / (CAPtestx - CAPtestm)$$

If no graphical or tabular data are provided for DFACT, then DFACT can also be estimated from a knowledge of the component's effective output resistance, which is often provided in the data manual's device characterization section:

Equation 6.4.1.7-3
ESTIMATING DFACT FROM COMPONENT OUTPUT RESISTANCE

DFACT = RO / 1.443

where RO = component effective output resistance, ohms

Another way to estimate DFACT is from the component's short-circuit current which is often provided on the data sheet:

Equation 6.4.1.7-4
ESTIMATING DFACT FROM COMPONENT
SHORT CIRCUIT CURRENT

DFACT = 0.5 * Vcc / Isc

where   Vcc = supply voltage at which Isc is measured
           Isc = device short-circuit current, amps

Finally, the following estimated min/max values can be used for devices within several commonly-used logic families:

Table 6.4.1.7-1
ESTIMATED MIN/MAX DFACT VALUES
FOR COMMON PART TYPES
Over Commercial/Military Temperature & Voltage Range

DFACT = delay time per load capacitance, sec/farad (ps/pf)

PART FAMILY	COM/MIL	TYPE	DFACTm	DFACTX
FACT	74	All	9.9	25.3
	54		8.2	28.6
ALS	74	Standard	24.9	63.4
	54		20.5	71.8
ALS	74	Driver	10.5	26.9
	54		8.7	30.4
AS	74	Standard	12.5	31.8
	54		10.3	36.0
AS	74	Driver	5.3	13.4
	54		4.3	15.2
FAST	74	Standard	18.6	47.4
	54		15.3	53.7
FAST	74	Driver	9.4	24.0
	54		7.8	27.2
HCMOS	74	Standard	31.2	81.8
	54		20.1	86.6
HCMOS	74	Driver	13.2	34.7
	54		9.0	39.5
LS	74	Standard	49.9	127.2
	54		41.1	144.1
LS	74	Driver	23.1	58.8
	54		19.0	66.6

EXAMPLE: Find the device delay for a 74AC241 bus driver IC driving a total load capacitance of 35.1 picofarads. The power supply Vcc for the circuit is +5V +/- 10%.

DDEL = DEL + DFACT * (Cload - Ctest)

From the data sheet for the 74AC241, the device delay DEL over its Vcc and temperature range is 1.0 to 7.5 ns for a test capacitance Ctest of 50 pf. For DFACT, since the data manual provides no graphs or tabular values for other load capacitances, Table 6.4.1.7-1 above is used. From the FACT/All/74 row, DFACT is estimated to be 9.9 to 25.3 seconds per farad.

DDELm = DELm + DFACTm * (Cload - Ctest)
     = 1.0E-9 + 9.9 * (35.1-50)E-12
     = 0.85 nanoseconds

DDELx = DELx + DFACTx * (Cload - Ctest)
     = 7.5E-9 + 25.3 * (35.1-50)E-12
     = 7.12 nanoseconds

6.4.1.8 Total Path Delay

The total delay for a given path is the sum of the intrinsic path delay plus the device delay:

Equation 6.4.1.8-1
TOTAL PATH DELAY

TDEL = PDEL + DDEL seconds

where   PDEL = intrinsic path delay (EQ 6.4.1.4-1)
         DDEL = device delay (EQ 6.4.1.7-1)

PDEL and DDEL are calculated for each clock and data path to obtain the total CLK and DAT values required for the timing margins analysis per EQ 6.4.1.1-1 or EQ 6.4.1.1-2.

EXAMPLE: Recompute the earlier timing margins analysis using the min/ max delay values, including the path and device parameters discussed above.

PATH DATA & CALCULATIONS

<———————Known From Design————————> <———————Calculated———————————>

Using Design Equations

Data	PL	Path Mils	Ccomp	LO	CO	CLF	Zo	TPDA	PDEL	PCAP	Cload
Path	in	W H T	pf	nh/in	pf/in		ohms	ns/in	ns	pf	pf
U1 to U2	2	10 20 1.5	4.5	12.66	1.54	1.57	90.6	0.223	0.477	3.08	7.6
U2 to U3	3	10 20 1.5	4.5	12.66	1.54	1.40	90.6	0.200	0.600	4.62	9.1

Clock	PL	Path Mils	Ccomp	LO	CO	CLF	Zo	TPDA	PDEL	PCAP	Cload
Path	in	W H T	pf	nh/in	pf/in		ohms	ns/in	ns	pf	pf
U4 to U3	14	10 15 1.5	13.5	12.66	1.54	1.27	90.6	0.182	2.542	21.6	35.1

DEVICE DATA

<————————Known From Design—————————> <———Calculated————>

Using Design Equations

Ref	Part No.	Test Cap	DFACT min	max	DEL nanoseconds min	max	DDEL nanoseconds min	max
U1	74AC74	50 pf	9.9	25.3	2.5	10.5	2.08	9.43
U2	74AC241	50 pf	9.9	25.3	1.0	7.5	0.60	6.47
U4	74AC241	50 pf	9.9	25.3	1.0	7.5	0.85	7.12

U3 Required setup time = -1.0 to 3.0 ns
U3 Required hold time = -3.5 to 0.5 ns

Note that in the DDEL calculations, the min/max DEL values are used in conjunction with the min/max DFACT values; i.e., it is assumed that slower devices will also have more delay due to load capacitance (correlated parameters).

DELAY CALCULATION RESULTS PER PATH:

Data Paths	TDEL	
	nanoseconds	
	min	max
U1 to U2	2.53	9.87
U2 to U3	1.20	7.07

Clock Path
U4 to U3	3.39	9.87

Therefore,

CLK= delay of U4 = 3.39 to 9.66 ns

DAT= delay of U1 + U2 = 2.53 + 1.20 to 9.87 + 7.07 = 3.72 to 16.94 ns

Setup Margin mSUm = CLKm - DATx + PERm - OFFx - tSx
$$= 3.39\text{-}16.94\text{+}10\text{-}0\text{-}3$$
$$= \text{- }6.55 \text{ ns}$$

Hold Margin mHDm = DATm - CLKx + OFFm - tHx
$$= 3.72\text{-}9.7\text{+}0\text{-}0.5$$
$$= \text{- }6.44 \text{ ns}$$

Sample Timing Margins Results Including Effects of All Delay Variables

Record # 1: mSU ======================= Status: Alert

ALERT probability = .2668295 %

Setup margin, ns	Minimum	Average	Maximum
Specification............	0		50
Actual.....................	-6.54502	5.19884	16.9427

The Design Analysis Handbook

```
    |                    x                    |
    |                    x                    |
    L                   xxx                   L
    I                  xxxxx                  I
    M                 xxxxxxx                 M
    m                xxxxxxxxx                x
    |              xxxxxxxxxxxxxx             |
 xxxxxxxxx|xxxxxxxxxxxxxxxxxxxxxxxxxxxxxxxxxxxxxxxx  |
___^_____^____
    M                                        M
    I                                        A
    N                                        X
```
Distribution weight = 8.95, centering = .50
Standard Deviation = 1.596167

————————SOLUTIONS————————

mSU = CLK - DAT + PER - OFF - tS

= ((tU4+((9.9+(tU4-1.0)*15.4/6.5)*1E-3)*(35.1-50))+2.542)-
((tU1+((9.9+(tU1-2.5)*15.4/8)*1E 3)*(7.650))+0.447+(tU2+
((9.9+(tU2-1.0)*15...

MIN = (3.39449)-(16.93951)+(10)-(0)-(3) = -6.54502

MAX = (9.66503)-(3.72233)+(10)-(0)-(-1) = 16.9427

Record # 2: mHD =======================Status: Alert
ALERT probability = .2317286 %

Hold margin, ns	Minimum	Average	Maximum
Specification............	0		50
Actual......................	-6.4427	5.30116	17.04502

```
       |                    x                    |
       |                    x                    |
       L                   xxx                   L
       I                  xxxxx                  I
       M                 xxxxxxx                 M
       m                xxxxxxxxx                x
       |               xxxxxxxxxxx               |
xxxxxxxxx|xxxxxxxxxxxxxxxxxxxxxxxxxxxxxxxxxxxxxxxxxxxx |
____^_____^____
```

```
  M                                        M
  I                                        A
  N                                        X
```

Distribution weight = 8.95, centering = .50
Standard Deviation = 1.596167

─────────────SOLUTIONS─────────────

mHD = DAT - CLK + OFF - tH
    = ((tU1+((9.9+(tU1-2.5)*15.4/8)*1E-3)*(7.6-50))+0.447+(tU2 +
       ((9.9+(tU2-1.0)*15.4/6.5)*1E-3)*(9.1-50))+0.600)-
       ((tU4+((tU4-1.0)*15....
MIN = (3.72233)-(9.66503)+(0)-(.5) =-6.4427
MAX = (16.93951)-(3.39449)+(0)-(-3.5) = 17.04502

6.4.2 Controlling Waveform Distortion

Distortion of the digital waveform can be created by *transmission line effects*, particularly when using "high-speed" logic. Severe distortion can cause device malfunction or even destruction. As some designers have found to their dismay, if transmission line effects are ignored during the design, the hardware may not function even though the design is otherwise flawless.

Since the edge speeds (output rise and fall times) of ICs are independent of the clock rate, the fast transitions of high-speed logic families can create waveform and timing distortion due to transmission line effects even at low clock rates.

CONTROLLING WAVEFORM DISTORTION TIP #1
To minimize waveform distortion problems,
use the slowest digital logic family that
will meet timing requirements.

The above tip is also good for reducing EMI.

Using high-speed logic requires a detailed knowledge of printed wiring board characteristics in addition to a thorough understanding of logic device drive characteristics. The key PWB and device parameters have already been discussed in the previous section; this section contains the equations to use to avoid transmission line troubles.

By considering driver characteristics and proper adjustments of path impedances and delays, the maximum allowable unterminated path lengths and the correct values of resistances for both series and parallel terminations can be determined. You can also validate the capability of line drivers to drive terminated lines.

6.4.2.1 Maximum Length of Unterminated Line

A printed-wiring path is a transmission line which can be approximately modeled as sections of inductance and capacitance as shown below:

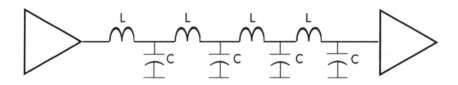

Figure 6.4.2.1-1
LC Approximation of Printed Wiring Path

The waveform distortion effects of the LC sections can be ignored when the device driving the path has output transition (rise and fall) times which are large compared to the delay time of the path. A general rule for the maximum allowable unterminated line length PLx is

Equation 6.4.2.1-1
MAXIMUM UNTERMINATED LINE LENGTH

$$PLx = tT/(2*TPDA)$$

where
tT = path driver output transition time
$TPDA$ = adjusted characteristic path delay/length (EQ 6.4.1.3-3)

If the maximum path length is used as calculated above, a peak undershoot or overshoot of approximately 15% of the logic swing will occur.

EXAMPLE: What is the maximum unterminated path length for a 4009 CMOS hex inverter buffer operating on a +5V supply? Assume that the device is being used to drive the 14-inch microstrip path used in the earlier examples:

The 4009 data sheet shows a minimum output transition time (50 pF load) of 35 nS.

$PLx = tT/(2*tPDA)$
$= 35ns/(2*0.182 \text{ ns/in})$
$= 96.4$ inches, or about 8 feet

Therefore, transmission line effects are in general not noticeable on circuits employing relatively slow edge speeds, such as 4000-series CMOS logic.

EXAMPLE: For the same microstrip path, what is the maximum unterminated path length for a 74AC74 flip-flop operating on a +5V supply?

The 74AC74 data sheet does not provide rise/fall times, but the manual indicates that the rise/fall times for the family are "rated" at 3 ns, so let's use 70% of that as an estimate of the fastest edge speed:

$$PLx = tT/(2*tPDA)$$
$$= 2.1ns/(2*0.182 \ ns/in)$$
$$= 5.8 \ inches$$

Therefore, when using higher-speed logic, transmission line distortion effects can start to be significant for short path lengths.

EXAMPLE: What are the maximum unterminated path lengths for the timing margins example in the previous section? Should these paths be terminated?

Device	Path	tT Edge Speed (Estimated min)	tPDA ns/in	PLx inches	PL inches	Termination Required?
74AC74	U1-U2	2.1 ns	0.223	4.70	2	NO
74AC241	U2-U3	2.1 ns	0.200	5.25	3	NO
74AC241	U4-U3	2.1 ns	0.182	5.78	14	YES

If you wish to *avoid* having to terminate a line, you can try the following:

CONTROLLING WAVEFORM DISTORTION TIP # 2
TO AVOID TERMINATIONS:
- *Keep the path length below PLx*
- *Use a slower driver (increase tT)*
- *Reduce the capacitive gate load (will decrease TPDA)*

The device output transition time tT is generally not provided by the manufacturer except for "typical" or "family" characteristics, which for high-speed logic devices are in the range of 5 or less nanoseconds. Such general data may understate the edge speed capability of a specific device, depending upon the part type, its capacitive loading, and the transition time of the input test waveform. Therefore the worst case (fastest) driver transition time should be estimated from minimum loading, maximum driver speed, and a fast waveform at the driver's input. If you need to estimate the fastest edge speed, you can use the following formula:

Equation 6.4.2.1-2
ESTIMATING THE FASTEST EDGE SPEED

$$tT = 2.2*ROM*COUTm$$

where ROM   = minimum driver effective output resistance, ohms
            = 1.443*DFACTm
               where DFACTm = minimum device capacitive drive
                                          capability, delay time per load
                                          capacitance (EQS 6.4.1.7-2 to 6.4.1.7-4
                                          or Table 6.4.1.7-1)

   COUTm = minimum output capacitance close to driver output

The above equation does not include the edge speed limitation inherent to internal device operation, so the calculated tT will be a smaller (faster edge speed) than actual, which will result in a shorter max path length PLx than is actually required.

Since the transition time at the start of a transmission line will not be reduced by capacitive loading further down the line, COUTm is the device output capacitance plus any stray capacitance located in the near vicinity of the output. This will typically be in the vicinity of 10 to 20 picofarads.

EXAMPLE: What is the estimated output transition time for a 74AS1004 hex inverter buffer? Assume a minimum nearby capacitance of 15 pf.

From Table 6.4.1.7-1, the minimum drive factor DFACTm is estimated to be 5.3 ps/pF, therefore

ROM = 1.443 *DFACTm
        = 1.443 *5.3
        = 7.65 ohms

$$tT \quad = 2.2*ROM*COUTm$$
$$= 2.2*7.65*15 \; E\text{-}12$$
$$= 0.25 \; nS$$

EXAMPLE: What is the maximum unterminated path length for the 74AS1004 device, assuming an adjusted characteristic delay of 0.22 ns/in?

$$PLx \quad = tT/(2*tPDA)$$
$$= 0.25ns/(2*0.22ns/in)$$
$$= 0.6 \; inches!$$

Although the formula for tT is overly conservative, we can still guess that the maximum line length is on the order of an inch. This example helps to explain why transmission-line problems are encountered in high-speed logic more often that might be expected based upon typical transition times and unadjusted characteristic delay data. For example, using the 74AS1004's typical DFACT value of 9.35 ps/pf and test load of 50 pF, the value for tT = 1.48 nS. If no gate loading were applied to the line, then tPDA = tPD = 0.142 nS/in, and PLx = 1.48/(2*0.142) = 5.21 inches. Therefore, it is apparent that the line length values seen in IC applications literature are often not suitable as high-speed guidelines, because (1) devices can be much faster than typical, and (2) even a moderate amount of gate loading will greatly increase the delay/length of a path.

6.4.2.2 Terminating the Line

To prevent waveform distortion on lines which must be longer than the max unterminated length PLx, *series* or *parallel* resistive terminations are employed.

For the series method, the terminating resistor is placed at the output of the line driver. For the parallel method, the terminating resistor is placed from the receiving device's input to ground. Frequently a DC blocking capacitor will be used in series with the parallel termination resistor as shown to eliminate DC power dissipation.

The series method of termination can only be used when the receiving devices are all clustered at the far end of a single path (no branches). This is because an undistorted waveform only occurs at the end of the path. At other points along the path, the waveform will not generally be useable.

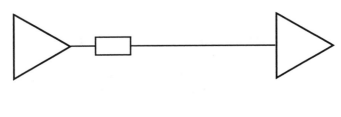

Figure 6.4.2.2-1
Series Termination

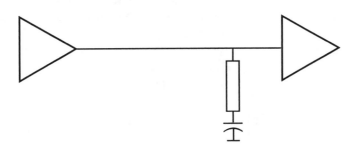

Figure 6.4.2.2-2
Parallel Termination

The parallel (or shunt) termination is usually preferred because it results in an acceptable waveform at any point on the path. To avoid the complexities of handling differing path impedances along the line, parallel paths should have no branches. A disadvantage of parallel-terminated paths is that they are harder to drive, and for some drivers the series termination may prove to be the only practical alternative.

For both types of terminations, the value of the terminating resistance is related to the *adjusted* path impedance, *not* the characteristic impedance Zo:

CONTROLLING WAVEFORM DISTORTION TIP # 3
Termination resistor values should be based on the ADJUSTED path impedance, NOT the characteristic impedance.

Equation 6.4.2.2-1
ADJUSTED PATH IMPEDANCE

$$Zoa \ (parallel) \ = Zo/CLF$$
$$Zoa \ (series) \quad = 2*Zo/(CLF+1)$$

where Zo = path characteristic impedance, ohms (EQ 6.4.1.2-1, 6.4.1.2-2)
   CLF = path capacitance load factor (EQ 6.4.1.3-2)

For a parallel termination, the required value of terminating resistance is equal to Zoa (parallel). For the series termination, the value of terminating resistance RTERM is equal to Zoa (series) *minus* the driver's output impedance RO:

Equation 6.4.2.2-2
VALUES OF TERMINATING RESISTANCE

$$RTERM \ (parallel) = Zoa \ (parallel)$$
$$RTERM \ (series) \quad = Zoa \ (series) - RO$$

where Zoa = adjusted path impedance, ohms (EQ 6.4.2.2-1)
   RO = driver equivalent output impedance, ohms
   = 1.443*DFACT
      where DFACT = device capacitive drive capability, delay time per load capacitance (EQS 6.4.1.7-2 to 6.4.1.7-4 or Table 6.4.1.7-1)

EXAMPLE: What is the value of terminating resistance for the 14-inch path from the previous example, assuming that the path can be series or parallel terminated?

RTERM (parallel) = Zoa = Zo/CLF
$$= 90.6/1.27$$
$$= 71.1 \text{ ohms}$$

For the series case,

RO = 1.443 * 17.6 (where 17.6 ps/pf is average DFACT value from Table 6.4.1.7-1)
$$= 25.4 \text{ ohms}$$

RTERM (series) = Zoa - RO = 2*Zo/(CLF+1) - RO

RTERM (series) = 2*90.6/(1.3 + 1) - 25.4
$$= 79.7 - 25.4$$
$$= 54.3 \text{ ohms}$$

6.4.2.3 Checking the Line Driver

After calculating the value for the terminating resistance, don't forget to check the driver's capability to drive the terminated line. In some cases, the adjusted path impedance is too low to be driven by the line driver!

CONTROLLING WAVEFORM DISTORTION TIP # 4
Always check to be sure that a line driver is capable of driving the adjusted impedance of its terminated line.

The above tip holds for *all* termination cases, even those where a blocking capacitor is used.

For the parallel terminated line, when the driver changes state its output sees an impedance equal to the path impedance Zoa. To check whether the driver can drive this impedance, calculate the voltage divider ratio created by the maximum driver output resistance and the adjusted line impedance:

Expression 6.4.2.3-1
LINE DRIVER CHECK for PARALLEL-TERMINATED LINE

RTERM/(RTERM+RO) must be greater than LOGIC SWING/VCC

where   RTERM = value of termination resistor (EQ 6.4.2.2-2)
                 RO = driver equivalent output impedance, ohms

EXAMPLE: Check the typical line driving capability of the 74AC241 device used for the 14-inch path from the prior example, assuming a parallel termination:

RTERM for the parallel-terminated line was previously calculated to be 71.1 ohms. RO (typical) for the 74AC241 was previously estimated to be 25.4 ohms.

$$\text{RTERM/(RTERM+RO)} = 71.1/(71.1 + 25.4)$$
$$= 0.74$$

The 74AC family requires a logic swing of approximately 67% of the supply voltage, so the worst case 74% ratio calculated above is satisfactory.

EXAMPLE: Check the *worst case* line driving capability of the 74AC241 device, assuming a parallel termination:

The maximum drive factor for the 74AC241 is estimated to be 25.3 ps/pf (from Table 6.4.1.7-1), therefore

$$ROX = 1.443 * 25.3 = 36.5 \text{ ohms}$$

$$\text{RTERM/(RTERM+ROX)} = 71.1/(71.1 + 36.5)$$
$$= 0.66$$

The worst case 66% ratio calculated above is marginal with respect to the required 67% logic swing ratio.

For the series terminated line, if the value calculated for RTERM is less than zero, then the output impedance of the driver is too large to match the line, so a series terminator cannot be used:

Expression 6.4.2.3-2
LINE DRIVER CHECK FOR SERIES-TERMINATED LINE

RTERM must be greater than ZERO

What happens if the driver can't drive the line? Here are a few things to try:

CONTROLLING WAVEFORM DISTORTION TIP # 5
IF THE DRIVER CAN'T DRIVE THE TERMINATED PATH:

- Use a faster driver (lower RO)
- Reduce the capacitive loading (increases the adjusted impedance Zoa)
- Increase the path length (counterintuitive, but works for large capacitive loads because the loads will have less of an effect on the path's characteristic impedance Zo)
- Increase the path's characteristic impedance (requires a change in the path geometries or PWB dielectric)

6.4.2.4 Effect of Terminations on Path Delays

The use of terminations will affect the characteristic delay TPD, with a parallel termination (or unterminated line) having less delay than a series terminated path:

Equation 6.4.2.4-1
ADJUSTED PATH DELAY

TPDA (unterminated or parallel terminated) = TPD*CLF
(same as basic Equation 6.4.1.3-3)

TPDA (series terminated)   = TPD*(2*CLF-1)

where   TPD = characteristic path delay, time/length (EQ 6.4.1.3-1)
                  CLF = capacitive load factor (EQ 6.4.1.3-2)

EXAMPLE: Find the adjusted path delay of the 14-inch line from the previous examples, assuming a parallel termination:

$$TPDA = TPD*CLF$$
$$TPD = 0.142 \text{ ns/in}$$
$$CLF = 1.27$$
$$TPDA = 0.142 * 1.27$$
$$= 0.182 \text{ ns/in}$$

For the 14-inch line, the intrinsic path delay PDEL is therefore

$$PDEL = TPDA * PL \text{ (EQ 6.4.1.4-1)}$$
$$= 0.182 * 14$$
$$= 2.54 \text{ ns}$$

EXAMPLE: Find the adjusted path delay of the 14-inch line from the previous examples, assuming a series termination:

$$TPDA = TPD*(2*CLF-1)$$
$$TPDA = 0.142*(2*1.27-1)$$
$$= 0.221 \text{ ns/in}$$

and the intrinsic path delay PDEL is

$$PDEL = 0.221 * 14$$
$$= 3.09 \text{ ns}$$

6.4.3 Miscellaneous Timing Tips

MISCELLANEOUS TIMING TIPS

- Avoid using capacitors in series with digital outputs for pulse-forming purposes. The capacitive discharge can create negative spikes which may cause the receiving IC to malfunction or even be damaged.

- At a digital boundary where asynchronous inputs enter a synchronous system, there is no way to guarantee that a timing violation will not occur. However, there are techniques which can reduce the probability of occurrence to an acceptably small number. For more information on this potential *metastability* problem, see "Exorcise Metastability from Your Design," David Shear, *EDN*, December 10, 1992.

- Don't forget that LSTTL data sheet parameters are only specified for nominal voltage and temperature; i.e., you'll need to estimate min/max values for analysis purposes.

- Although a small MOSFET is often used as a convenient means of level-shifting or inverting an incoming logic waveform, care should be exercised for critical timing applications. If the threshold voltage of the FET is not close to the mid-point of the input waveform's transitions, and the input has rise or fall times which are significant compared to its period, pulse width distortion will occur.

- In addition to transmission-line waveform distortion discussed earlier, a similar but unrelated form of distortion can be created at high-speed IC outputs when their package lead plus path inductances react with the path and load capacitances to create an LC ringing response, characterized by an undershoot on a falling edge and an overshoot on a rising edge. This waveform ringing distortion is called "ground bounce" although Vcc can be involved as well as ground. Attention to ground-bounce specifications and applications data provided by the IC vendor will mitigate this form of distortion.

6.5
GROUNDING AND LAYOUT

Proper grounding and layout are essential for correct circuit operation. Unlike discrete circuit elements such as resistors, capacitors, ICs, etc., grounding/layout is a set of distributed or physical parameters, such as ground planes, wire routing, component placement, and controlled impedances. Many standard rules of thumb can be employed by the designer, but these are not appropriate for all applications. It is better to really understand the nature of the grounding/layout problem, which will arm the designer with the knowledge required to create effective solutions. This understanding comes from an appreciation of a few key unifying concepts as described in this section.

6.5.1 The Signal Loop

Many designers look at signal paths as originating at a driver output and terminating at a receiver input. To understand planes and other grounding/ layout parameters, however, one should envision an entire signal loop :

KEY TIP FOR PROPER GROUNDING and LAYOUT
Look at all signal paths as loops

A signal must go from a power source or local reservoir (capacitor), down a signal path to a load, through the load, and back through a return path to the driving device and power source. Some basic circuit loops are shown below:

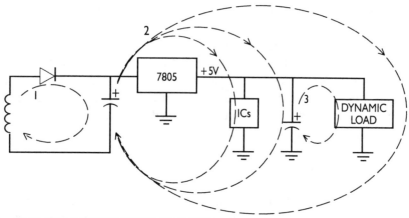

1 = LOCAL HIGH PULSE CURRENT PRIMARY POWER LOOP
2 = LOAD CURRENT LOOPS
3 = LOCAL DYNAMIC LOAD CURRENT LOOP

Figure 6.5.1-1
Basic Circuit Loops

The primary goal of signal transmission is to get the voltage from the source to the load with minimal distortion. Therefore, the signal path and its return must have a very low impedance with respect to the load. For DC signals, the concern therefore is that the path resistances be adequately small; for higher frequency signals, path inductances must also be adequately small. For digital signals, the paths also often require a controlled impedance to prevent waveform distortion due to transmission-line effects. (Also see Section 6.4, "Digital Applications.")

Additional signal transmission goals are to minimize pickup from other signals or noise sources, and to minimize interference with other circuits. By visualizing the basic signal current loop, one can begin to understand how to best satisfy these sometimes conflicting signal transmission goals.

6.5.2 Low Impedance

For low impedance the power source path, signal path, and return path should be as short as practical, and have low resistance and inductance.

The power source for a digital driver or comparator may be a local filter capacitor; the capacitor provides the current required to drive the path

capacitance during switching transitions. Hence local decoupling capacitors should have short, low inductance leads; surface mount devices are
ideal. (Also see 6.5.6 "Decoupling.")

A common example of a signal path is the printed wiring path, with the
path dimensions selected to meet impedance goals. Although the return
path is often a plane, it doesn't *have* to be, provided that the return path has
a low enough impedance. In fact, as will be discussed below, an improp-
erly implemented plane can even degrade performance.

6.5.3 Controlled Impedance

High speed digital and video designs often require controlled impedance paths. Such paths can be created by the signal path *and its return*. A
printed wiring path of given dimensions over a ground plane yields such a
controlled impedance, provided that the ground plane *is not interrupted*
beneath the signal path by cutouts due to segmenting or component place-
ments:

MAINTAINING A CONTROLLED IMPEDANCE
Never run a controlled-impedance path
over a cut in the ground plane

A controlled impedance path depends upon the repeating parameters
of capacitance and inductance per unit length, like a lumped-element model
of a transmission line. As a signal travels down a path over a plane, all is
well unless all of a sudden these characteristics abruptly change. A cut in
a ground plane will force the return current to diverge and spread around
the cut, like an ocean wave hitting a rock. The resulting "splash" will not
only disrupt the path impedance, creating significant waveform distortion
(ringing, over/undershoot, etc.), but will also likely generate appreciable
radiated noise.

6.5.4 Preventing Pickup

The integrity of a current loop is disturbed by (a) allowing the loop to
be commingled with other loops, (b) capacitive coupling of nearby paths to
the loop, or (c) electromagnetic radiation impinging on the loop.

Preventing Commingling

To prevent commingling, an individual return path can be provided for each loop:

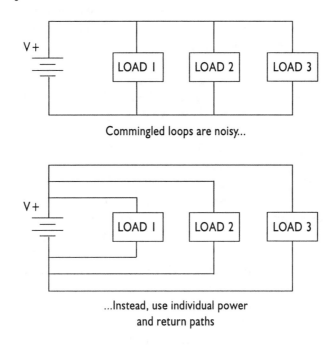

Commingled loops are noisy...

...Instead, use individual power
and return paths

Figure 6.5.4-1
Combined vs. Separate Circuit Loops

Because of layout constraints, separate returns cannot always be implemented without tortuous routing which will cause capacitive coupling problems. Therefore a ground plane is often employed as a universal return path for a group of signals. It is important to realize that the plane is employed not because a plane is "good," but because the plane is a solution to a circuit density problem.

When several loops are commingled as a result of sharing a common plane, problems can arise because currents from the various loops are now contaminating each other's loops. The contamination is proportional to the impedance of the shared portion of the plane and the current amplitudes of the various signals. Even though a plane has a low impedance, high frequency and/or high amplitude signals can still create enough voltage drop across the plane to disturb sensitive circuitry. Therefore, *plane segmenting* is recommended to isolate signal groups. The plane is physically cut into two or more segments, or islands, above which the circuit groups are placed. The planes are connected together at a point of minimal area where all loops may "touch" but share very little common path length; typically this occurs at the power supply return.

AVOIDING GROUND PLANE SIGNAL CONTAMINATION
Segment the ground plane

By grouping high-level and low-level circuits above their own ground segments, signal loops associated with each circuit group are not commingled. This prevents the high-level signals from disturbing the low-level signals. Of course, signals within a group are still commingled, so if there are any "wild" signal loops within a group, they should not use the ground plane, but should use individual return paths. Examples of an isolated loop are: a printed wiring signal path with a similar return path which ideally runs directly underneath (or above) the signal path; a twisted pair line; or a coax cable.

Since segmenting a plane requires a cut, the ground-cut problem discussed earlier can be encountered if one is not careful. This hazard is often encountered in analog-digital or digital-analog circuits, where designers try to segment sensitive analog circuits from the more-noisy digital circuits. Many vendors of A/D or D/A ICs separate their ICs into a "digital" side and an "analog" side, with separate ground return points for each side. Although this facilitates ground segmenting, if the design contains more than one of these chips (not uncommon), then it is often very difficult to ensure that the paths all stay above their own segments; i.e, one may be tempted to violate the don't-cross-a-ground-cut rule. But *don't do it!* Instead, provide any intra-segmented paths with their own individual return paths, or route the paths around the segment at the common connect point for the planes.

Preventing Capacitive Pickup

Signal contamination due to capacitive coupling is generally controlled by keeping paths short and by using segmented ground planes. These guidelines are listed below, along with some added tips for very sensitive paths:

PREVENTING CAPACITIVE PICKUP

In general,

- keep the paths short
- use ground segmenting

For very sensitive paths,

- physically isolate the path
- use a differential driver/receiver for the path (cancels the coupled signal)
- use a twisted pair for the path and return
- use a shielded cable for the path and return (or shield the entire sensitive circuit)

Preventing Electromagnetic Pickup

Electromagnetic coupling from a nearby circuit (e.g., a switchmode power supply) or from the ambient environment (e.g., 60-cycle hum from power equipment) can contaminate the current loops by injecting or inducing error voltages in the loops.

When the coupling is primarily due to electric fields, such as power supply noise generated by high-voltage low-current switching waveforms, the coupling mode will be capacitive, and the tips in the previous section can be employed.

When the coupling is primarily due to magnetic fields, such as 60-cycle hum created by power line current flowing through input power leads, use the options below:

PREVENTING MAGNETIC PICKUP

- Keep the loop area as small as possible:
  - keep path lengths short
  - keep the path and return as close together as possible:
    - signal path over plane
    - signal path over individual return
    - use a twisted pair for the path and return
    - use a coax cable for the path and return

- Use a differential driver/receiver for the path (cancels the coupled signal)

Shielding magnetic fields is more difficult than shielding electric fields, so shielding is not listed as a preferred option. If shielding is necessary, then a high-permeability shield (e.g., mumetal), will probably be required. (Also see Section 6.6, "EMI / Noise Control.") The coax cable option above helps guard against magnetic pickup by virtue of the cable's small effective loop area, not its shield.

6.5.5 Preventing Interference

The same measures used to protect sensitive circuits from pickup can be used to prevent noisy circuits from interfering with other circuits:

PREVENTING INTERFERENCE

For noisy circuits,

- keep the high-amplitude or high-speed loop areas as small as possible:
  - keep path lengths short
  - keep the path and return as close together as possible:
    - signal path over plane
    - signal path over individual return
    - use a twisted pair for the path and return
    - use a coax cable for the path and return

– 170 –

- physically isolate the circuit from other circuits (separation)

- place the circuit on its own ground plane segment

- place the noisy circuit within a shielded enclosure

Note that "twisted pair" above refers to any signal and return pair, not just digital signal twisted-pair lines. For example, one of the most effective ways of reducing low-frequency hum interference is by twisting together power line leads.

6.5.6 Decoupling

In addition to their role in maintaining the low impedance of a signal loop, local decoupling capacitors are also essential for preventing malfunctions or oscillations in ICs. Although experienced design engineers are well aware of this, sometimes decoupling capacitors get moved during layout away from the ICs they're supposed to filter. This may be because on some schematics all the decoupling capacitors are shown on a separate page, which doesn't give the layout person a clue as to the importance of their physical location. Although this can be caught during layout check, why not reduce the chance of rework and draw the capacitors on the schematic in the location where they're needed; e.g., right across an op amp or comparator's power terminals?

There are several rules of thumb for local decoupling, but device peculiarities prevent a "universal" solution. For example, some ICs require decoupling capacitors in unlikely spots, some require a ceramic and an electrolytic, some require a damping resistor in series with a capacitor, etc. Therefore, to be safe it's essential to refer to the fine print in the data sheet or application note provided by the part vendor, and use the recommended decoupling.

6.6
EMI / NOISE CONTROL

The control of Electromagnetic Interference (EMI) is often described as an art rather than engineering. While it's true that EMI is a richly complex and diverse area of engineering, the few key concepts and equations presented in this section will help the non-specialist obtain pretty good estimates for these basic EMI questions:

1. How much EMI can my circuit pick up?
2. How much EMI can my circuit generate?

Topic (1) addresses EMI susceptibility , while topic (2) addresses EMI interference. The concepts and techniques used for one topic are directly related to the other. For additional information, see Section 6.5, "Grounding and Layout" and the References and Recommended Reading section.

6.6.1 Some Key EMI Concepts

The concepts described in this section will help the electronics design engineer visualize the EMI problem from a practical circuit-effects perspective. Using these concepts, the designer will be able to sidestep some common EMI misconceptions, and avoid related EMI hazards which often slip unknowingly into a design.

Field Characteristics

The two components of an electromagnetic wave are the *electric field* (e-field) and the *magnetic field* (h-field). When observed at a distance approximately greater than the wavelength of the field y divided by 2*Pi, the distance is called the *far-field* distance, and the field has a unified electric/magnetic characteristic impedance of 377 ohms. Far fields (or planewaves) can be readily shielded (typically more than a factor of 10,000, or > 80 dB) by non-magnetic (copper, aluminum) or magnetic (steel or mumetal) materials. Shielding is effective over the entire frequency spectrum of concern for most electronics circuits (from DC to more than 10 GHz).

EMI TIP #1
Shielding is Very Effective For Far-Field EMI

To calculate the far-field,

Equation 6.6.1-1
FAR-FIELD DISTANCE VERSUS FREQUENCY

Dff >= y/(2*Pi) meters

where   y = wavelength
             = c/f, meters
             c = speed of light = 3E8 meters/sec
             f = frequency, Hz

Table of Dff versus frequencies from 10 Hz to 1 GHz

Frequency, Hz	10	100	1000	10000	1e5	1e6	1e7	1e8	1e9
Dff, meters	4.8e6	4.8e5	47746	4775	477.5	47.75	4.775	0.477	0.048

EXAMPLE: An AM broadcast station 1/2 mile distant is transmitting at 1
MHz. Is the broadcast signal a plane-wave?

1/2 mile = 0.5*5280 ft x 0.3047 meters/ft
         = 804 meters

Dff >= 3E8/(2*Pi*f) meters
    >= 3E8/(2*Pi*1E6)
    >= 47.7 meters

Since the distance must be greater than 47.7 meters away for
1 MHz and the source is 804 meters away, the signal will be
a plane wave.

EXAMPLE: A power supply operating at 50 KHz is located 12 inches away from your sensitive analog circuitry. Will the noise generated by the power supply have a far-field characteristic? (Consider harmonics up to 250 KHz.)

12 inches x 0.02539 meters/in
= 0.305 meters

Dff(250KHz) >= 3E8/(2*Pi*250E3)
>= 191 meters

Since the distance must be greater than 191 meters away for 250 KHz and the source is 0.305 meters away, the noise will *not* have far-field characteristics.

For frequencies around 10 MHz or lower, the far-field threshold is approximately 5 meters or more; i.e., for most circuit-to-circuit or even rack-to-rack interference, the fields of concern will *not* be far-field for a large part of the spectrum.

When the noise-to-circuit distance is less than the far-field transition point Dff, the range is called the *near-field*. As the noise-source distance moves from Dff to less than Dff, the field begins to assume the character of an electric field or a magnetic field, depending on the nature of the noise source. Using the 377 ohm plane-wave impedance as a baseline, the e-field impedance *increases* x10 whereas the h-field impedance *decreases* x10 for every x10 decrease in distance. For example, at a distance of 0.1*Dff, the e-field has an impedance of 3.8 Kohms, and the h-field has an impedance of 38 ohms. Therefore one of the primary characteristics of the two fields is their impedance: the e-field has a high impedance (Ze = higher voltages / lower currents) and couples onto circuitry capacitively, while the h-field has a low impedance (Zh = lower voltages / higher currents) and couples onto circuitry inductively:

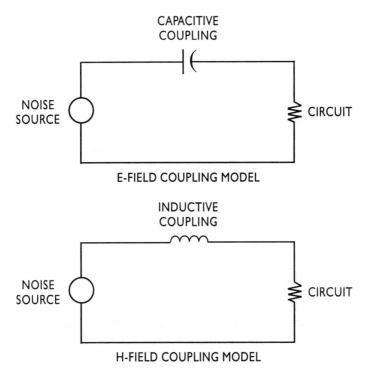

Figure 6.6.1-1
Simple E-Field and H-Field Coupling Models

Since circuits operate by definition with current flowing through various paths and their returns (current loops), much near-field noise will be of the h-field variety. A typical h-field is created by AC line current flowing through a power supply input and return leads. In circuit areas where high frequency waveforms charge/discharge local stray capacitances, significant e-field noise may also be present. A typical e-field is generated by the sharp voltage waveform on the collector of a power transistor in a switchmode power supply.

The models for e-field and h-field coupling shown in the figures above are duals of each other; therefore circuit techniques which work for mitigating one type of coupling may not work at all for the other type of coupling.

EMI TIP #2
Know Thy Near-Fields

For example, for an e-field noise source a circuit impedance can be lowered to reduce the noise pickup, since this will cause more of the noise to be "dropped" across the capacitive coupling of the source. However, if the source is an h-field, lowering the circuit impedance won't help too much, especially at lower frequencies, since the source impedance is very low. By extending this impedance argument to the use of a shield, an important conclusion is obtained for those who may be tempted to look upon shielding as a panacea:

EMI TIP #3
Shielding May Be Ineffective For Low-Frequency H-Fields

Since h-field coupling is inductive, the coupling impedance will be lower at low frequencies, allowing more of the noise voltage to couple to a circuit. Even if a shield is placed around the circuit, the noise will generate shield currents which in turn will generate noise fields inside the shield, making the shield appear somewhat transparent, particularly at low frequencies. For example, an aluminum shield will only provide on the order of x10 reduction in noise levels for frequencies in the vicinity of 100 KHz, and very little attenuation of noise frequencies less than 10 KHz. (This is one reason why 120 Hz rectified power line "hum" can be a difficult EMI challenge.) One way to stop the shield penetration is to use high permeability shielding material, such as steel or mumetal. A high-perm magnetic shield will *absorb* part of the magnetic field's energy, provided that the selected material does not saturate for the maximum expected field strength. However, one of the best means for coping with magnetic fields is by reducing the *loop area*, as discussed in the following section.

For e-fields, the coupling impedance is high compared to a shield impedance, which prevents most of the noise energy from coupling to the shield. As the model suggests, at higher frequencies e-field coupling improves, which allows more noise coupling to the shield. Fortunately however, at the higher frequencies a shield will absorb most of the incident energy, resulting in good performance for all frequencies.

Loop Area

The degree of noise received/transmitted is proportional to the loop area of the conductive path doing the receiving/transmitting:

EMI TIP #4
Identify The Loop Areas

A loop is usually visualized as something with a circular shape, but as referred to herein it means any closed current path. Such paths are often rectangular (path above ground return), but can actually be a straight dimension such as a cable or rod, where the balance of the "loop" is through the capacitance associated with the cable or rod. There are generally two loop area types of concern, *differential-mode* and *common-mode*. Differential-mode refers to the voltage on a signal path measured with respect to the signal return path; common-mode refers to a voltage on *both* the signal and return paths measured with respect to a "common" reference, often earth ground. Examples of some common loops are shown below:

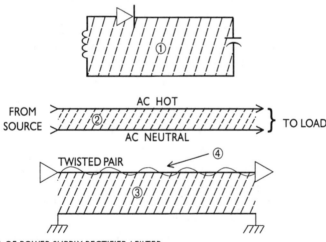

① LOOP AREA OF POWER SUPPLY RECTIFIER / FILTER
② LOOP AREA OF AC POWER LEADS
③ LOOP AREA (COMMON-MODE) OF SIGNAL PAIR ABOVE GROUND
④ LOOP AREA (DIFFERENTIAL-MODE) OF SIGNAL LEAD TWISTED
   WITH RETURN LEAD (AREA BETWEEN LEADS)

Figure 6.6.1-2
Examples of Loop Areas

6.6.2 Estimating EMI Pickup

How much noise can your circuit pick up? Some formulas and tips for estimating *narrowband* noise (noise from a single frequency or a closely-spaced group of frequencies) are presented below. They are followed by a methodology and related equations for estimating *broadband* noise (noise generated across a wide band of frequencies, as created by pulse wave-forms).

Narrowband Pickup

Equation 6.6.2-1
A ROUGH ESTIMATE FOR EMI PICKUP

Volts (pickup) = Ef*LD

where   Ef = noise source electric field strength, volts/meter
              LD = maximum linear dimension of exposed circuit path

EXAMPLE: What is the rough estimate for the EMI pickup in a circuit which has an exposed 6-inch path which must operate in a noise e-field of 10 volts/meter?

Volts (pickup) = 10 volts/meter x 6 inches x 0.02539 meters/inch
= 1.52 volts

The actual value of EMI pickup will depend on the loop configuration and dimensions, the spectrum of the noise source, the circuit's bandwidth and sensitivity, plus other factors. Nonetheless, a simple calculation using the above equation gives one a rough feel for the significance of circuit dimensions versus field strengths.

For a variety of man-made sources (radars, broadcast stations, pagers), the field strength 1 KM from the source can range roughly from tens of millivolts/meter to hundreds of volts/meter. At the 1KM range a 6-inch exposed "antenna" may pick up unwanted noise amplitudes in the range of millivolts to tens of volts. Noise estimates for different distances from the source can be obtained by noting that the far-field strength is proportional to the reciprocal of the distance. For example, for a transmitter which generates an e-field at 1KM of 5 V/M, the e-field at 0.1KM will be 5*1KM/ 0.1KM = 50 V/M.

In the general case, the pickup path will not be exposed all by itself, but the return path will also be exposed. If the return path is immediately adjacent to the outgoing path, then the voltage induced on both paths will be almost the same and of the same polarity, so the net induced *differential-mode* voltage between outgoing terminal and return terminal will be very small. However, the outgoing and return terminals will both have the same induced *common-mode* voltage, which may still couple into other circuitry.

If the return path is not very close to the signal path, the loop area which is formed will pick up ambient noise proportional to the loop area. For e-field cases, the pickup voltage Vpu induced in series with a current loop is:

Equation 6.6.2-2
E-FIELD EMI PICKUP

$Vpu = Ef*1.414*L*SIN(2.22*H/y)$ volts

where   Ef = electric field strength, volts/meter
     Loop = approximate rectangle shape of dimensions H x L:

     H = loop height above reference, meters
     L = loop length, meters

Limit the value of H, L to a maximum of 1/2 the wavelength:

     H = H   if H <= y/2
     H = y/2   if H > y/2
     L = L   if L <= y/2
     L = y/2   if L > y/2
     y = wavelength
       = 3E8/f
     f = frequency, Hz

The above equation is very powerful and can be used to estimate the noise pickup for a wide variety of situations.

The Design Analysis Handbook

EXAMPLE: What is the noise pickup on a circuit loop which consists of a 8-inch path running 1/4 inch above an 8-inch return path (differential-mode noise pickup), with the loop optimally positioned with respect to a noise e-field of 7 volts/meter at a noise frequency of 60 MHz?

Loop = H x L = 1/4 in x 8 in

Ef = 7 V/M

H = 1/4 in x 0.02539 m/in
   = 6.35E-3 meters

L = 8 in x 0.02539 m/in
   = 0.203 meters

y = 3E8/60E6
   = 5.0 meters

y/2 = 2.5 meters

H,L < y/2

Vpu = Ef*1.414*L*SIN(2.22*H/y)
    = 7*1.414*0.203*SIN(2.22*6.35E-3/5)
    = 5.67 millivolts

EXAMPLE: For the previous example, what is the noise pickup on *both* the signal and return paths (common-mode noise pickup) with respect to chassis (earth) ground which is 5 inches away?

Loop = H x L = 5 in x 8 in

Ef = 7 V/M

H = 5 in x 0.02539 m/in
   = 0.127 meters

L   = 8 in x 0.02539 m/in
    = 0.203 meters

y   = 5.0 meters
y/2 = 2.5 meters

H,L < y/2

Vpu = Ef*1.414*L*SIN(2.22*H/y) volts
    = 7*1.414*0.203*SIN(2.22*0.127/5)
    = 113.3 millivolts

EXAMPLE: What is the common-mode noise pickup from the previous
example if the noise source has a frequency of 1GHz?

H = 0.127 meters
L = 0.203 meters
y = 3E8/1E9
   = 0.30 meters

y/2 = 0.15 meters

H < y/2

L > y/2, use y/2 for L

Vpu = Ef*1.414*L*SIN(2.22*H/y)
    = 5*1.414*0.150*SIN(2.22*0.127/0.3)
    = 856.1 millivolts

For h-field cases, the pickup voltage Vpu induced in series with a current loop is:

Equation 6.6.2-3
H-FIELD EMI PICKUP

Vpu = Bfd*4.243E4*L*SIN(2.22*H/y) volts

where Bfd = magnetic flux density, Gauss

Loop = approximate rectangle shape of circuit pickup
dimensions H x L:

H = loop height, meters
L = loop length, meters

Limit the value of H, L to a maximum of 1/2 the wavelength:

H = H   if H <= y/2
H = y/2   if H > y/2
L = L   if L <= y/2
L = y/2   if L > y/2
y = wavelength
= 3E8/f
f = frequency, Hz

EXAMPLE: A switching power supply is generating a 400 KHz h-field
with a flux density of 350 milliGauss measured at a sensitive
circuit. The circuit has an input path length of 5.4 inches located 1/32 inch above a ground plane. What is the noise
voltage pickup in the circuit's input path?

Loop = H x L = 1/32 in x 5.4 in

Bfd = 350E-3

H = 1/32 in x 0.02539 m/in
= 0.793E-3 meters

L = 5.4 in x 0.02539 m/in
= 0.137 meters

y = 3E8/400E3
= 750 meters

y/2 = 375 meters

H,L < y/2

Vpu = Bfd*4.243E4*L*SIN(2.22*H/y) volts
= 350E-3*4.243E4*0.137*SIN(2.22*0.793E-3/750) volts
= 4.78 millivolts

Broadband Pickup

In many cases the noise source will have a *broadband* characteristic; i.e., it will generate a pulsed e-field, whose Fourier frequency components will cover a broad spectrum. The approach discussed above can be expanded to handle such cases:

PROCEDURE FOR CALCULATING PULSED
E-FIELD NOISE PICKUP

1. Calculate the spectral density of the e-field pulse.

2. Use Equation 6.6.2-2 (e-field) or 6.6.2-3 (h-field) to calculate the volts for a range of sample frequencies across the pulse spectrum, or volts/Hz.

3. Integrate the results of step (2) to obtain the peak noise volts.

This is easier than it sounds. First, the pulse spectrum is calculated from the following relationship:

Equation 6.6.2-4
PULSE SPECTRAL DENSITY

Psd = 2*Fpk*tP*((1/(1+f*(Pi*tP)))*(1/(1+f*(Pi*tR))))) volts(amps)/m/Hz

where   Fpk = field peak pulse amplitude, volts/meter (e-field) or
                  amps/meter (h-field)
            f = frequency, Hz
           tP = pulse width, sec (measured at 50% of pulse height)
           tR = pulse rise time, sec (measured between 10% and 90%
                  of pulse height)

When Psd is plotted against frequency, the curve will have a low-frequency amplitude of 2*Fpk*tP, with a first breakpoint at f1 = 1/(Pi*tP)
and a second at f1 = 1/(Pi*tR):

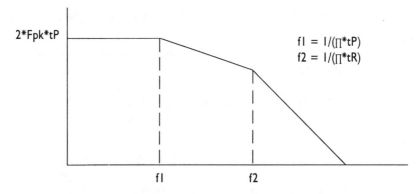

Figure 6.6.2-1
Fourier Spectrum of Pulse, dB

EXAMPLE: What is the spectral density of an e-field pulse with a peak
                amplitude of 20 V/M, a risetime of 130 nanoseconds, and a
                pulse width of 15 microseconds?

                Fpk = 20 V/M
                tP = 15E-6 sec
                tR = 130E-9 sec

$$Psd = 2*Fpk*tP*((1/(1+f*(Pi*tP)))*(1/(1+f*(Pi*tR))))$$
$$= 2*20*15E-6*((1/(1+f*(Pi*15E-6)))*(1/(1+f$$
$$*(Pi*130E-9)))) \text{ V/M/Hz}$$

f	Psd
Hz	V/M/Hz
1000	5.728e-4
3162	5.215e-4
10000	4.062e-4
31623	2.379e-4
100000	1.009e-4
316228	3.342e-5
1000000	8.852e-6
3162278	1.745e-6
10000000	2.499e-7
31622777	2.892e-8
100000000	3.042e-9
316227766	3.09e-10
1000000000	3.11e-11
3162277660	3.12e-12
10000000000	3.12e-13

A plot of Psd in dB (20*LOG(Psd)) is shown below:

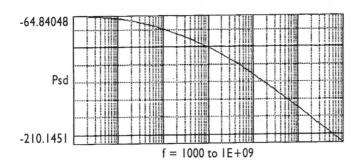

Figure 6.6.2-2
Plot of Pulse Spectral Density, dB Volts/Meter/Hz

After obtaining Psd, the noise voltage density is obtained by evaluating Equation 6.6.2-2 across the frequency range, substituting Psd (V/M/Hz) for Ef (V/M). The result (identified as broadband pickup volts, or Vpub) will be in volts/Hz, instead of volts (Vpu).

Assume that the pulse e-field is applied to a 34-inch path running 12 inches above its return:

H = 12 in x 0.02539 m/in = 0.3047 meters

L = 34 in x 0.02539 m/in = 0.8633 meters

f   where   H = y/2 = 4.92e8 = 492 MHz

f   where   L = y/2 = 1.74e8 = 174 MHz

Since Vpub is being evaluated for several frequencies, the values of L and H must be compared to 1/2 the wavelength (1/2 x 3E8/f) for each frequency. Vpub is calculated at each frequency, using Psd and the appropriate values for H and L. In the tabulation below, the rightmost column is the approximate integral of Vpub over each decade, obtained by taking the value of Vpub at the decade intermediate frequency (10 dB intermediate point = factor of 316) and multiplying by the difference between the decade frequencies bracketing the sample point; e.g., the value for the decade from 1MHz to 10 MHz = 1.519E-8*(10E6-1E6) = 0.137 volts. The integral is taken for each decade, and the results summed to obtain the total integral across the spectrum, or a noise pulse of 515 millivolts.

f Hz	Psd V/M/Hz	y/2 Meters	H Meters	L Meters	Vpub V/Hz	Integral V/Dec	20LOG(Psd)
1000	5.728e-4	150000	0.3047	0.8633	1.576e-9		-64.840483
3162	5.215e-4	47434.165	0.3047	0.8633	4.539e-9	0.000	-65.654728
10000	4.062e-4	15000	0.3047	0.8633	1.118e-8		-67.826041
31623	2.379e-4	4743.4165	0.3047	0.8633	2.07e-8	0.002	-72.473078
100000	1.009e-4	1500	0.3047	0.8633	2.778e-8		-79.921016
316228	3.342e-5	474.34165	0.3047	0.8633	2.908e-8	0.026	-89.520976
1000000	8.852e-6	150	0.3047	0.8633	2.437e-8		-101.05875
3162278	1.745e-6	47.434165	0.3047	0.8633	1.519e-8	0.137	-115.16227
10000000	2.499e-7	15	0.3047	0.8633	6.878e-9		-132.04444
31622777	2.892e-8	4.7434165	0.3047	0.8633	2.515e-9	0.226	-150.77727
100000000	3.042e-9	1.5	0.3047	0.8633	8.e-10		-170.33562
316227766	3.09e-10	0.4743416	0.3047	0.4743	1.36e-10	0.122	-190.19124
1000000000	3.11e-11	0.15	0.15	0.15	5.91e-12		-210.14509
3162277660	3.12e-12	0.0474342	0.0474	0.0474	1.87e-13	0.002	-230.13044
10000000000	3.12e-13	0.015	0.015	0.015	5.92e-15		-250.12581

Total noise volts = 0.515

Although math software can be used to obtain more precise integration, the simple spreadsheet tabulations used in the above example give a good approximation.

Note that peak noise volts calculated above is for a single pulse. For a repeating pulse waveform, you can obtain the average noise volts by multiplying the noise calculated above times the duty cycle. In general, however, good EMI design requires immunity to single peak transients, so the above procedure can be used to estimate the maximum generated peak instantaneous noise for either single transient pulses or repetitive waveforms.

EXAMPLE: Repeat the previous example for a 10 MHz clock pulse waveform with a risetime of 5 nanoseconds and a duty cycle of 50%.

The 10 MHz clock rate has a period of 100 nanoseconds; for a 50% duty cycle the pulse width is 50 nanoseconds:

Fpk = 20 V/M
tP = 50E-9 sec
tR = 5E-9 sec

f Hz	Psd V/M/Hz	y/2 Meters	H Meters	L Meters	Vpub V/Hz	Integral V/Dec
1000	2e-6	150000	0.3047	0.8633	5.5e-12	
3162	2e-6	47434.165	0.3047	0.8633	1.74e-11	0.000
10000	2e-6	15000	0.3047	0.8633	5.5e-11	
31623	1.99e-6	4743.4165	0.3047	0.8633	1.73e-10	0.000
100000	1.97e-6	15000	0.3047	0.8633	5.41e-10	
316228	1.9e-6	474.34165	0.3047	0.8633	1.65e-9	0.001
1000000	1.7e-6	150	0.3047	0.8633	4.684e-9	
3162278	1.27e-6	47.434165	0.3047	0.8633	1.108e-8	0.100
10000000	6.72e-7	15	0.3047	0.8633	1.85e-8	
31622777	2.24e-7	4.7434165	0.3047	0.8633	1.947e-8	1.753
100000000	4.66e-8	1.5	0.3047	0.8633	1.271e-8	
316227766	6.61e-9	0.4743416	0.3047	0.4743	2.902e-9	2.612
1000000000	7.6e-10	0.15	0.15	0.15	1.44e-10	
3162277660	7.9e-11	0.0474342	0.0474	0.0474	4.76e-12	0.043
10000000000	8e-12	0.015	0.015	0.015	1.53e-13	

Total noise volts = 4.508

The methodology above can be further exploited for more complex situations by working in the frequency domain before integrating. For example, the noise density Vpub on an incoming wire can be applied to a filter in the frequency domain by inserting the appropriate columns (component impedances and filter response function). After obtaining the filter output in the frequency domain, the integrating technique as demonstrated above is performed to get the filtered pulse amplitude.

For another common example, the noise density can be applied to the shield impedance of a cable and the frequency-domain shield current can be obtained. This current is multiplied by the transfer impedance of the cable to obtain the voltage density on the interior wires (core) of the cable. This is finally integrated to obtain the noise pulse on the cable core.

EMI TIP #5
HOW TO CALCULATE COMPLEX PULSE EFFECTS
Work in the frequency domain using the methodology of this section, then perform the final integration to obtain the resultant pulse response.

6.6.3 Reducing EMI Pickup

The first line of defense is to keep sensitive circuits confined to small, compact areas, located away from noisy circuits (also see Section 6.5, "Grounding and Layout").

For additional protection from far-field and e-field noise sources, shielding will offer orders of magnitude of noise reduction, if the shield has no holes or seams large enough to allow the noise frequencies to enter. When a hole or seam is present, if its maximum dimension is equal to or greater than approximately 1/2 the wavelength of the frequency of interest, then the shield's effectiveness is severely diminished:

EMI TIP #6
HOW TO MAKE AN EMI SHIELD VANISH
*Put a hole or a slot in the shield with a maximum
dimension equal to or greater than 1/2
the wavelength of the frequency of interest.*

For frequencies lower than 1/2 the wavelength, the shield attenuates the frequency by a factor of 10 per frequency decade (20 dB attenuation per decade below the "cutoff"):

Equation 6.6.3-1
ATTENUATION OF A SHIELD WITH AN APERTURE

AAf = Aperture Attenuation Factor = $1E5*(1+LN(L/W))/(L*f*1E-3)$

where
LN = natural logarithm function
W = aperture maximum width, meters
L = aperture maximum length, meters

Limit the value of W, L to a maximum of 1/2 the wavelength:

W = W   if W <= y/2
W = y/2   if W > y/2
L = L   if L <= y/2
L = y/2   if L > y/2

$$y = \text{wavelength}$$
$$= 3E8/f$$
$$f = \text{frequency, Hz}$$

EXAMPLE: A circuit is located in a shielded enclosure which is exposed to an external e-field noise frequency of 100 MHz. A shielded cable is routed through the enclosure via a slot which is 4" long and 1/8" high. What is the attenuation of the noise? (Ignore the presence of the cable and consider the slot alone.)

$$1/2\ y = 0.5*3E8/100E6$$
$$= 1.50 \text{ meters}$$

$$W = 0.125 \text{ inch x } 0.02539 \text{ meters/inch}$$
$$= 3.17 \text{ millimeters} << y/2$$

$$L = 4.0 \text{ inches x } 0.02539 \text{ meters/inch}$$
$$= 102 \text{ millimeters} << y/2$$

$$AAf = 1E5*(1+LN(L/W))/(L*f*1E-3)$$
$$= 1E5*(1+LN(102/3.17))/(102E-3*100E6*1E-3)$$
$$= 43.8$$

Therefore the 100 MHz noise will be attenuated by a factor of 43.8, or 20*LOG(43.8) = 32.8 dB. For example, an exterior 100 MHz noise amplitude of 1 volt will be attenuated to 1/43.8 = 23 millivolts inside the enclosure. Since an unslotted shield would have a much larger attenuation factor (thousands to tens of thousands), the slot leakage is significant.

EXAMPLE: A high-speed µP circuit board generates e-field noise at switching harmonics in the vicinity of 180 MHz which severely interfered with sensitive analog circuitry on a nearby PWB. Measurements determined that the noise needed to be reduced by a factor of 2000 (66 dB), so the µP circuitry was placed in a plastic enclosure with metallic plating, which achieved 86

dB attenuation in the frequency range of interest. During installation of the circuit, a 2-inch long by 6-mil wide scratch was accidentally made in the plating.
How did the scratch affect the shielding?

1/2 y = 0.5*3E8/180E6
    = 0.83 meters

   W = 0.006 inch x 0.02539 meters/inch
     = 0.152 millimeters << y/2

   L = 2.0 inches x 0.02539 meters/inch
     = 50.8 millimeters << y/2

AAf = 1E5*(1+LN(50.8/0.152))/(50.8E-3*180E6*1E-3)
    = 74.5
    = 37.4 dB

i.e., the scratch degraded the shielding effectiveness to well below requirements.

Equation 6.6.3-1 does not include the effects of the aperture depth, which can increase the attenuation factor. The extra attenuation is a function of the hole dimensions and is approximated by:

Equation 6.6.3-2
EFFECT OF APERTURE DEPTH

ADf = Aperture Depth Factor = 10^(1.5*D/L)

where   D = depth of aperture
               L = length of aperture

The Design Analysis Handbook

EXAMPLE: A 1/4" hole has a depth of 1/16 inch. What is the extra attenuation which will be gained compared to a hole with negligible depth?

$$ADf = 10^{(1.5*D/L)}$$
$$= 10^{(1.5*0.0625/0.25)}$$
$$= 2.37$$
$$= 20*LOG(2.37) = 7.50 \text{ dB}$$

EXAMPLE: What if the hole depth is increased to 1/4 inch for the previous example?

$$ADf = 10^{(1.5*0.25/0.25)}$$
$$= 31.6$$
$$= 20*LOG(31.6) = 30.0 \text{ dB}$$

EXAMPLE: The required attenuation of an enclosure is 50 dB at 1 GHz. The selected metallic enclosure has 80 dB of shielding effectiveness (10,000 attenuation) at 1 GHz. It is discovered that a 1/4" diameter hole must be placed in the enclosure. What must the hole depth be to ensure that the shielding effectiveness is no less than 60 dB (1000 attenuation)?

$$1/2 \text{ y} = 0.5*3E8/1E9$$
$$= 150 \text{ millimeters}$$

$$W = L = 0.25 \text{ inch x } 0.02539 \text{ meters/inch}$$
$$= 6.35 \text{ millimeters} << y/2$$

$$AAf = 1E5*(1+LN(6.35/6.35))/(6.35E-3*1E9*1E-3)$$
$$= 15.7$$

The attenuation with the hole without depth considered is 15.7 and the requirement is 1000, so the extra attenuation required is 1000/15.7 = 63.7. Therefore,

$$ADf \text{ (required)} = 63.7$$

Solving for D,

$20*LOG(ADf) = 20*(1.5*D/L) = 20*LOG(63.7) = 36.1$
$D = 36.1*L/(20*1.5) = 36.1*6.35E-3/30$
    $= 7.64$ millimeters
    $= 0.301$ inches

To check, the total attenuation is

$AAf*ADf = 1E5*(1+LN(L/W))/(L*f*1E-3)*10^{\wedge}(1.5*D/L)$
  $= 1E5*(1+LN(6.35/6.35))/(6.35E-3*1E9*1E-3)*$
    $10^{\wedge}(1.5*7.64/6.35)$
  $= 1005$
  $= 60.04$ dB

As mentioned earlier, it can be difficult to achieve low-frequency h-field shielding. Sometimes shielded/coax cables help, but this is due to a reduction in loop area, not the shield itself; i.e., the shield acts as the return path for the signal current. This assumes the return current flows through the shield and not through an alternate path such as a signal ground to which both ends of the shield are connected:

EMI TIP #7
ENSURING MINIMAL LOOP AREA IN A CABLE
*Make sure that the return current is forced to flow on a path adjacent to the signal current, and **not** through an alternate path such as a ground plane or chassis ground.*

GUIDELINES FOR REDUCING EMI PICKUP

1. Reduce the circuit's LOOP AREAS:
- minimize path lengths
- minimize path separation:
  - run the signal over a dedicated return path
  - use a twisted pair
  - use a coax cable

2. Circuit PLACEMENT:
    • locate sensitive paths away from noise sources
    • orient sensitive paths for minimal coupling to the noise source

3. Add SHIELDING:
    • use shielded cables
    • use a shielded enclosure
    • use high-perm non-saturating shields for h-fields

6.6.4 Mitigating EMI Pickup

After taking steps to minimize noise coupling, additional measures can be employed to reduce the effect of the noise that does enter the system. This noise can be modeled as a voltage source in series with the circuit loop:

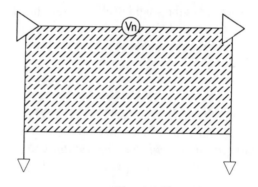

Figure 6.6.4-1
EMI Pickup = Noise Volts in Series with Loop

GUIDELINES FOR REDUCING THE EFFECTS OF EMI PICKUP

1. To reduce the effects of COMMON-MODE pickup:
   - isolate the circuit ground from earth ground:
     - float the circuit (DC and AC; requires minimal capacitive coupling from circuit to earth)
     - use opto or transformer isolators
     - add series impedance:
     - insert filters or ferrite beads in return paths
   - cancel the pickup:
     - use balanced pairs

2. To reduce of the effects of DIFFERENTIAL-MODE pickup:
   - filter the circuit interfaces
   - tailor the circuit bandwidth to its application

6.6.5 Estimating EMI Generation

How much noise is your circuit generating? To estimate the maximum noise from a wire carrying a current, use the following formulas:

Equation 6.6.5-1
ELECTRIC FIELD STRENGTH FROM A WIRE

$$Ef = 30*I*L/r^2*SQR((r*2*Pi*f/3E8)^2-1+1/(r*2*Pi*f/3E8)^2) \text{ V/M}$$

where
- I = current flowing in wire, amps
- L = length of wire, meters
- r = measurement point distance from wire, meters
- f = frequency of current, Hz
- SQR = square root function

Assumes $L \ll r$ and $L \ll y$, where y = wavelength = 3E8/f

The Design Analysis Handbook

Limit the value of L to a maximum of 1/2 the wavelength:

$$L = L \quad \text{if } L <= y/2$$
$$L = y/2 \quad \text{if } L > y/2$$

EXAMPLE: A switchmode power supply with a plastic cover employs a power
FET whose drain waveform contains a 5-amp pulse ringing at 6
MHz due to parasitic oscillations. The affected path is 6 inches
long (0.152 meter), including its return. What is the peak e-field
noise due to the oscillation at a measuring distance of 1 meter?

y = 3E8/6E6 = 50 meters
r = 1 meter
L = 0.152 meters

L << r, y

Ef = 30*I*L/r^2*SQR((r*2*Pi*f/3E8)^2-1+1/(r*2*Pi*f/3E8)^2) V/M
= 30*5*0.152/1^2*SQR((1*2*Pi*6E6/3E8)^2-1+1/(1*2*Pi*6E6/3E8)^2)
= 180 volts/meter peak

Equation 6.6.5-2
ELECTRIC FIELD STRENGTH FROM A LOOP

Ef = 8E-7*I*A*(Pi*f)^2/(3E8*2*r)*SQR(1+(3E8/(2*Pi*f*r))^2) V/M

where   I = current flowing in loop, amps
A = area of loop, square meters
r = measurement point distance from wire, meters
f = frequency of current, Hz
SQR = square root function

"Loop" can be rectangular area with max dimension = D meters

Assumes D << r and << y, where y = wavelength = 3E8/f

Limit the value of D to a maximum of 1/2 the wavelength:

$$D = D \quad \text{if } D <= y/2$$
$$D = y/2 \quad \text{if } D > \ y/2$$

EXAMPLE: A high-frequency buffer drives a 3-inch jumper wire which is 1 inch above a ground plane. The peak current is 100 ma. What is the e-field noise at a measuring distance of 1 meter?

y = 3E8/50E6 = 6 meters
r = 1 meter
A= 3 in x 2 in = .0762 m x .0508 m = 3.87E-3 square meters

Dmax = .0762 m

D << r, y

$$Ef = \ 8E\text{-}7*I*A*(Pi*f)^2/(3E8*2*r)*SQR(1+(3E8/(2*Pi*f*r))^2)$$
$$= 8E\text{-}7*0.1*3.87E\text{-}3*(Pi*50E6)^2/(3E8*2*1)*SQR(1+$$
$$(3E8/(2*Pi*50E6*1))^2)$$
$$= 0.0176 \text{ volts/meter}$$

Equation 6.6.5-3
MAGNETIC FIELD STRENGTH FROM A WIRE

$$Hf = I*L*f/(3E8*2*r)*SQR(1+(3E8/(2*Pi*f*r))^2) \text{ amps/meter}$$

where
I = current flowing in wire, amps
L = length of wire, meters
r = measurement point distance from wire, meters
f = frequency of current, Hz
SQR = square root function

Assumes L << r and L << y, where y = wavelength = 3E8/f

Limit the value of L to a maximum of 1/2 the wavelength:

$$L = L \quad \text{if } L <= y/2$$
$$L = y/2 \quad \text{if } L > y/2$$

EXAMPLE: What is the magnetic field noise generated by the switchmode power supply from the earlier example at a distance of 1 foot?

$$y = 3E8/6E6 = 50 \text{ meters}$$
$$r = 1 \text{ foot} = 0.3048 \text{ meters}$$
$$L = 0.152 \text{ meters}$$

$$L < r, << y \quad (\text{approximate, since L not} << r)$$

$$Hf = I*L*f/(3E8*2*r)*SQR(1+(3E8/(2*Pi*f*r))^2)$$
$$= 5*0.152*6E6/(3E8*2*.3048)*SQR(1+(3E8/(2*Pi*6E6 *.3048))^2)$$
$$= 0.6515 \text{ amps/meter peak}$$

Equation 6.6.5-4
MAGNETIC FIELD STRENGTH FROM A LOOP

$$Hf = 1.67E-9*f*I*A/r^2*SQR((r*2*Pi*f/3E8)^2-1+1/(r*2*Pi*f/3E8)^2) \text{ A/M}$$

where
I = current flowing in loop, amps
A = area of loop, square meters
r = measurement point distance from wire, meters
f = frequency of current, Hz
SQR = square root function

"Loop" can be rectangular area with max dimension = D meters

Assumes D << r and << y, where y = wavelength = 3E8/f

Limit the value of D to a maximum of 1/2 the wavelength:

$$D = D \quad \text{if } D <= y/2$$
$$D = y/2 \quad \text{if } D > y/2$$

EXAMPLE: What is the magnetic field noise generated by the high frequency amplifier from the earlier example at a distance of 1
foot?

$$y = 3E8/50E6 = 6 \text{ meters}$$
$$r = 1 \text{ foot} = 0.3048 \text{ meters}$$
$$A = 3 \text{ in } x \text{ 2 in} = .0762 \text{ m } x .0508 \text{ m} = 3.87E\text{-}3 \text{ square meters}$$

$$Dmax = .0762 \text{ m}$$

$$D << r, << y$$

$$
\begin{aligned}
Hf &= 1.67E\text{-}9*f*I*A/r^{\wedge}2*SQR((r*2*Pi*f/3E8)^{\wedge}2\text{-}1+1/ \\
&\quad (r*2*Pi*f/3E8)^{\wedge}2) \\
&= 1.67E\text{-}9*50E6*0.1*3.87E\text{-}3/.3048^{\wedge}2*SQR((.3048*2* \\
&\quad Pi*50E6/3E8)^{\wedge}2\text{-}1+1/(.3048*2*Pi*50E6/3E8)^{\wedge}2) \\
&= 1.04 \text{ milliamps/meter}
\end{aligned}
$$

After estimating a noise field by using one of the above equations, the
effect on a circuit can be obtained by using Equation 6.6.2-2 (noise volts
due to e-field) or 6.6.2-3 (noise volts due to h-field). In the latter case, the
h-field noise calculated with Equation 6.6.5-3 or 6.6.5-4 is given in amps/
meter and needs to be converted to Gauss:

$$Gauss = 4*Pi*E\text{-}3*h$$
$$= 0.01257*h$$

EXAMPLE: What is the noise volts generated by the previous example at
the 1-foot distance in a wire which is 6 inches long and 2
inches above its return path?

The Design Analysis Handbook

Gauss = 0.01257*h
       = 0.01257*1.04E-3
       = 13.1 microGauss

Using Equation 6.6.2-3,

Loop = H x L = 2 in x 6 in

Bfd = 13.1E-6 Gauss
H   = 0.0508 meters
L   = 0.152 meters
y   = 3E8/50E6 = 6 meters
y/2 = 3 meters

H,L < y/2

Vpu= Bfd*4.243E4*L*SIN(2.22*H/y) volts
     = 13.1E-6*4.243E4*0.152*SIN(2.22*0.0508/6) volts
     = 1.59 millivolts

6.6.6 Reducing EMI Generation

The tips used to reduce EMI pickup can be applied to reduce EMI generation:

GUIDELINES FOR REDUCING EMI GENERATION

1. Reduce the noisy circuit's LOOP AREAS:
    • minimize path lengths
    • minimize path separation:
       - run the signal over a dedicated return path
       - use a twisted pair
       - use a coax cable

2. Circuit PLACEMENT:
    - locate noisy sources away from sensitive paths
    - orient noisy sources for minimal coupling to sensitive paths

3. Add SHIELDING:
    - use shielded cables for noisy circuit interface lines
    - place the noisy circuit within a shielded enclosure
    - use high-perm non-saturating shields for h-fields

4. FILTER the interface connections of noisy circuits

6.7
CIRCUIT POTPOURRI

6.7.1 Op Amp Trouble Spots

The following comments refer to standard voltage-feedback op amps (as differentiated from wide-band current-feedback op amps).

Oscillations Due to Improper Decoupling or Layout

TIP #1 for AVOIDING OP AMP OSCILLATIONS
Follow the vendor's recommendations
for power supply decoupling and layout

Oscillations Due to Load Capacitance

Some modern op amps have increased tolerance to output load capacitance, making them attractive candidates for applications for which op amps were previously unsuitable, such as line drivers. But many standard op amps are prone to oscillations when driving capacitive loads of even moderate value. In fact, interconnect capacitance due to cables or long printed wiring paths may be high enough to cause instabilities, even if there are no discrete load capacitors used in the design.

TIP #2 for AVOIDING OP AMP OSCILLATIONS
Check the op amp's capability of driving any
load capacitance, including cable, printed wiring,
or other interconnect capacitance

If you're not using an op amp that has a defined capacitive load capability, it will be necessary to take added steps to ensure the op amp's stability, such as using the standard isolation circuit shown below, or employing other compensation techniques as recommended by the op amp vendor.

Equation 6.7.1-1
ESTIMATING COMPONENT VALUES FOR A STANDARD
OP AMP CAPACITIVE ISOLATION CIRCUIT

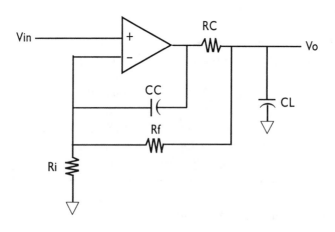

$RC = RO*Ri/Rf$ = compensation resistor

$CC = CL*RO/Rf*(1+Ri/Rf)^2$ = compensation capacitor

where   RO = op amp output impedance at high frequency, ohms
CL = load capacitance, farads

Because the equation above is an approximation, be sure to test the final circuit for stability.

EXAMPLE: What are the capacitive load compensation components required for the op amp circuit above with the following values?

Ri   = 10K
Rf   = 100K
Op amp = TLE2024
CL   = 0.05 μF
From the TLE2024 data sheet curves, the high frequency output impedance is approximately 34 ohms:

RO = 34 ohms

RC = RO*Ri/Rf
    = 34*10E3/100E3
    = 3.4 ohms (use 3.6 ohm standard value)

CC = CL*RO/Rf*(1+Ri/Rf)^2
    = 0.05E-6*34/100E3*(1+10E3/100E3)^2
    = 20.6 picofarads (use 22 pF standard value)

Oscillations Due to High-Valued Feedback Resistors

Another op amp oscillation hazard is the use of high-valued feedback resistors; e.g., 1 Megohm or thereabouts. With such high values, the parasitic capacitance at the op amp input now becomes significant, adding additional loop phase shift and reducing the stability margin.

Expression 6.7.1-2

CHECKING FOR OP AMP OSCILLATIONS
DUE TO HIGH-VALUED FEEDBACK

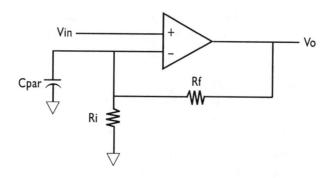

Make sure that: PMgc - PSfb > 45 degrees

where   PMgc = op amp phase margin at gain crossover, degrees

        PSfb = added phase shift due to feedback network
             = ATAN(fC/fP)

       ATAN = arctangent function

       fC = gain crossover frequency of op amp

       fP = frequency pole due to feedback + parasitic capacitance
           = 1/(2*Pi*Req*Cpar)

      Cpar = parasitic input capacitance
             (typically a few picofarads)

      Req = equivalent resistance of feedback network
           as seen from Cpar

The phase margin is the op amp data sheet value which assumes no extra phase shift from parasitic effects. The arctangent function computes the added phase shift caused by the feedback resistors and parasitic capacitance. After subtracting out the added phase shift, the phase margin for good stability should be greater than 45 degrees.

The phase margin of an op amp is obtained from the op amp data sheet by the following steps:

1. Find the curve showing the open loop gain versus frequency.

2. Locate on the vertical scale the closed loop gain. For example, for a closed loop gain of 10, find the 20 dB point.

3. Move horizontally to the right on the gain line until it intersects with the open loop gain curve; this is the gain crossover.

4. Move vertically down to find the crossover frequency fC.

5. If the op amp data sheet provides an open loop phase plot, find the phase margin (distance from 180 degrees total shift) at frequency fC.

6. If a phase plot is not available, estimate the phase margin by assuming that it will be approximately 45 degrees at the unity-gain frequency fU; i.e.,

$$\text{phase margin (est)} = \text{ATAN}(fU/fC)$$

EXAMPLE: Are oscillations likely for the op amp circuit above with the following values?

Ri   = 120K
Rf   = 1.2 Meg
Op amp = LM358

Assume that the parasitic capacitance at the input is 5 picofarads,

Cpar = 5E-12

From the LM358 data sheet, for the gain of 10 (20 dB) the gain crossover occurs at approximately 100KHz,

fC = 100 KHz

Req = parallel combination of Ri and Rf
    = 109K

fP = 1/(2*Pi*Req*Cpar)
    = 1/(2*Pi*109E3*5E-12)
    = 292 KHz

The data sheet doesn't provide phase data, so the phase margin at the gain crossover is estimated. From the data sheet, fU = approximately 800 KHz,

phase margin = ATAN(fU/fC)
             = ATAN(800/100)
             = 83 degrees

phase shift = ATAN(fC/fP) = ATAN(100/292)
            = 18.9 degrees

phase margin - phase shift = 83 - 18.9 = 64.1 > 45 degrees, so the circuit will be stable.

EXAMPLE: For the previous example, are oscillations likely if Ri is changed from 120K to 1.2 Megohm?

From the LM358 data sheet, for the gain of 2 (6 dB) the gain crossover occurs at approximately 600KHz,

fC = 600 KHz

Req = parallel combination of Ri and Rf
    = 600K

fP $= 1/(2*Pi*Req*Cpar)$
     $= 1/(2*Pi*600E3*5E-12)$
     $= 53$ KHz

phase margin $=$ ATAN(fU/fC)
                     $=$ ATAN(800/600)
                     $= 53$ degrees

phase shift $=$ ATAN(fC/fP) $=$ ATAN(600/53)
                  $= 85.0$ degrees

phase margin - phase shift $= 53 - 85.0 = -32.0$ which is NOT
> 45 degrees, so the circuit will be unstable.

   In such cases, the addition of a small cancelling capacitor CF across
the feedback resistor can prevent a problem:

TIP #3 for AVOIDING OP AMP OSCILLATIONS

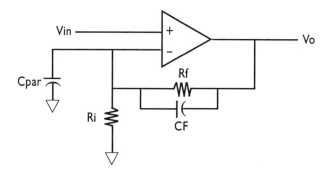

Avoid high-valued resistors in the feedback network.
If you can't, add a small capacitor CF across the
feedback resistor of the following value:

$$CF = Cpar*Ri/Rf$$

where Cpar = parasitic input capacitance

KEEPING THE OP AMP SPEEDY

Don't allow the op amp output to saturate:

- use clamping zeners or networks to clamp the inputs or outputs:

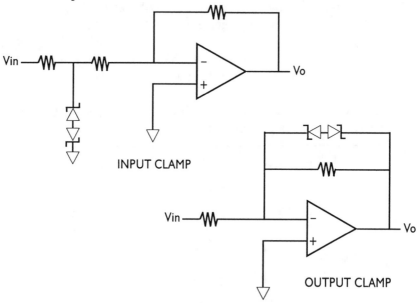

INPUT CLAMP

OUTPUT CLAMP

- when using clamps, watch out for:

  - zener capacitance effects on transient performance; use iso
    lation diode networks if speed is a concern.

  - max leakage: the clamp's max leakage should be negligible
    compared to circuit bias requirements.

  - zener clamps with low-noise designs: if biased near the knee,
    zeners can be very noisy; a zener clamp from output to input

can inject noise current into the op amp which will show up as an output noise voltage.

MAINTAINING OP AMP ACCURACY

• don't drive heavy loads with a precision op amp; the extra power dissipation will add temperature-induced error. Use a buffer inside the loop instead.

• keep the input signals within the specified common-mode range.

• pay attention to bias and offset current effects when using high-impedance networks.

6.7.2 Comparator Trouble Spots

Want to make a high-frequency burst oscillator, cheap? Just leave out the hysteresis resistors in your comparator circuits. This is also a good way to generate random interference.

AVOIDING COMPARATOR OSCILLATIONS

• use hysteresis

• follow the vendor's recommendations for power supply decoupling and layout

6.7.3 Power-On/Off, Reset, and Transients

If you assume that your circuit will always be operated in its linear range or within its common-mode ratings, sooner or later you will be proved wrong. Power-on/off ramps, reset and interrupt conditions, or input/output over-range transients will eventually kick your circuit into an "undefined" mode, which if not anticipated by the design may result in the activation of sneak circuit paths, output polarity reversals, latchup, etc. A moderate amount of preventive analysis and testing of initialization, overload, and transient conditions will help achieve a hearty fault-tolerant design. Note: When testing such conditions, *be careful*; you may get an unexpected and unpleasant surprise.

TIPS FOR ENSURING A FAULT-TOLERANT DESIGN

1. Assume that Bozo the Clown will be using your product.

2. Provide all power inputs with reverse-polarity and overvoltage protection. Use undervoltage lockout where appropriate.

3. Provide all power outputs with overvoltage protection and current limiting.

4. Provide controlled responses to system start-up, interrupt, and power-down.

5. Use isolation resistors, fault-tolerant ICs, or other hearty buffers in series with all signal inputs/outputs which interface with the "outside world."

6. Account for the effects of all controls set to all possible combinations over their full range.

7. Account for the effects of all possible on/off combinations of all power supplies.

8. Account for the effects of suddenly disconnecting each input and output.

9. Use small fuses or fusible resistors in series with key power paths to limit the damage caused by fault conditions.

One of the smallest and simplest, yet most treacherous designs is the reset circuit. Many reset circuits have an irreducible window of error, with the error dependent upon the ramp-up time of system power. For a solution, try a simple JFET as a reset device. A JFET is reliable and cost effective because of its passive (no power applied) low resistance, which for example can be used to initialize a soft-start capacitor in the discharged state. No other simple device provides this no-power shorting function.

USE A JFET FOR RELIABLE RESET

6.7.4 Timers

Timers such as the LM555 use one or two resistors and a capacitor as the time-setting elements. We've observed several instances where long time periods are attempted with one or two stages, requiring the use of very large values of R and C; e.g., megohms and tens or hundreds of microfarads.

To keep size within reasonable bounds, the capacitor type selected for such super-timers is generally an electrolytic; e.g., aluminum or tantalum. The problem with this is that electrolytics have non-negligible leakage currents, particularly when compared to the value of capacitor charge current. For example, for a total timing resistance of 1 megohm, the peak timing current may be on the order of 5 microamps. A capacitor leakage of only 1 microamp will therefore introduce a substantial timing error. Also, leakages can be much higher, particularly at high temperature (e.g., ten microamps). If the leakage is high enough, the leakage may drain the capacitor faster than it can be charged, resulting in a timer which never "times out":

AVOIDING THE INFINITE TIMER CIRCUIT

- ensure that capacitor and IC input leakages are negligible compared to the timing charge current

- use a digital countdown circuit for those really long timer applications

---

6.7.5 Frequency Response

You can generate a variety of frequency response curves with just a few handy equations:

Table 6.7.5-1
HANDY EQUATIONS FOR GENERATING
FREQUENCY RESPONSE CURVES

POLES and ZEROES:

POLEg = Pole gain, dB   = -20*LOG(SQR(1+(f/fP)^2))

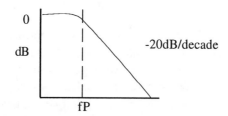

POLEp = Pole phase, degrees   = -ATAN2(1,f/fP)*180/Pi

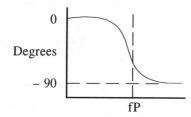

ZEROg = Zero gain, dB   = 20*LOG(SQR(1+(f/fZ)^2))

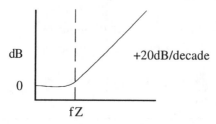

ZEROp = Zero phase, degrees $\quad\quad$ = ATAN2(1,f/fZ)*180/Pi

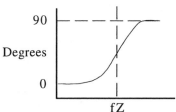

POLE2g = 2nd-order pole gain, dB
$\quad$ = -20*LOG(SQR((1-(f/fN)^2)^2+(2*dF*f/fN)^2))

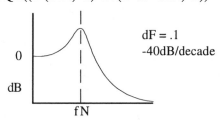

POLE2p = 2nd-order pole phase, degrees
$\quad$ = -ATAN2((1-(f/fN)^2),2*dF*f/fN)*180/Pi

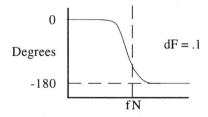

where $\quad$ f $\;$ = frequency, Hz
$\quad\quad\quad\quad$ fP $\;$ = pole frequency, Hz
$\quad\quad\quad\quad$ fZ $\;$ = zero frequency, Hz
$\quad\quad\quad\quad$ fN $\;$ = natural frequency, Hz
$\quad\quad\quad\quad$ dF $\;$ = damping factor
$\quad\quad\;$ ATAN2 $\;$ = 4-quadrant arctangent function, radians
$\quad\quad\;$ format $\;$ = ATAN2(real,imaginary)

EXAMPLE: What is the gain response of a network with independent poles
$\quad\quad\quad\quad$ at 1000 Hz, 3000 Hz, and an independent zero at 5000 Hz?

$$fRESPg = POLEg(1000) + POLEg(3000) + ZEROg(5000)$$
$$= -20*LOG(SQR(1+(f/1000)^2)) \quad POLEg(1000)$$
$$-20*LOG(SQR(1+(f/3000)^2)) \quad POLEg(3000)$$
$$+20*LOG(SQR(1+(f/5000)^2))dB \quad ZEROg(5000)$$

A plot of fRESPg is shown below:

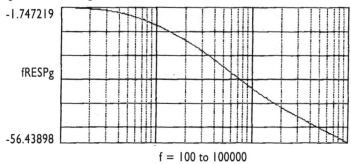

f = 100 to 100000

EXAMPLE: What is the phase response of the network from the previous example?

$$fRESPp = POLEp(1000) + POLEp(3000) + ZEROp(5000)$$
$$= -ATAN2(1,f/fP)*180/Pi \quad POLEp(1000)$$
$$= -ATAN2(1,f/fP)*180/Pi \quad POLEp(3000)$$
$$= +ATAN2(1,f/fZ)*180/Pi \ degrees \quad ZEROp(5000)$$

A plot of fRESPp is shown below:

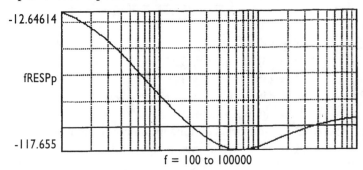

f = 100 to 100000

Filter functions can also be expressed as closed-form equations.

EXAMPLE: What is the gain response of the following Sallen Key low pass filter?

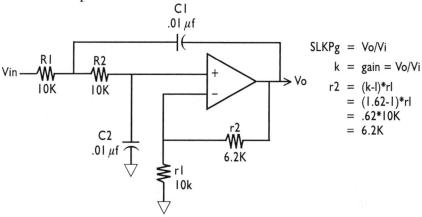

SLKPg = Vo/Vi
k = gain = Vo/Vi
r2 = (k-1)*rl
= (1.62-1)*rl
= .62*10K
= 6.2K

SKLPg = Sallen Key Low Pass gain, dB
= 20*LOG(A/SQR(B^2+C^2))

where   A = k/(R1*R2*C1*C2)
B = 2*Pi*f*(1/(R1*C1)+1/(R2*C1)+1/(R2*C2)-k/(R2*C2))
C = 1/(R1*R2*C1*C2)-(2*Pi*f)^2
f = frequency, Hz
k = gain = Vo/Vi

A plot of SKLPg is shown below:

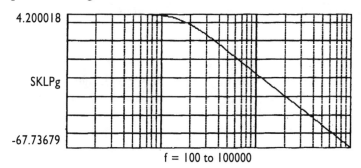

f = 100 to 100000

6.7.6 Miscellaneous Tips

Interfacing

Watch out for those boundaries between circuits on different power supplies; power sequencing problems are a common occurrence. The usual
scenario is when inputs of parts on an OFF power bus are connected to the
outputs of parts on an ON power bus. The OFF devices load down the ON
devices, causing them to malfunction. The ON supply may also be pulled
below its specified lower limit, causing system errors or shutdown. To avoid
this, do not rely on power switching to be "simultaneous"; instead place
ON devices in a high-impedance (e.g., tri-state) mode during the intervals
when receiving devices are unpowered. An alternative is to employ receiving circuits or devices which are designed to present a high impedance
to input voltages which exceed the receiver's supply voltage, even when
unpowered.

Protection from Test Equipment

Test equipment should contain circuitry to protect the unit under test; we've
seen more than one case where malfunctioning test equipment has damaged very expensive assemblies. Fuses alone are not completely effective
because they are slow to open, and may not protect against overvoltage
damage. Therefore, TPDs (Terminal Protection Devices) such as zener
diodes should also be employed across the power and data interfaces. Be
sure to use TPDs which are guaranteed to fail in a shorted condition.

Video Jitter

If the pixels are swimming in your new display design, it may be that a
nearby switching power supply isn't synchronized to a multiple of the scan
frequency.

Teflon Caution

Although Teflon (DuPont trademark) insulation has excellent electrical properties, we've observed cold flow problems when a Teflon-insulated cable was routed around a bracket or pressed against a ground plane. Mild mechanical pressure caused the insulation to separate over time, exposing the wire and causing a short. Perhaps some formulations have improved properties; unfortunately, the cable data sheets we've seen neglect to mention cold flow properties.

REFERENCES AND RECOMMENDED READING

GENERAL, ANALOG, POWER

Data Sheets and Application Notes for components you're using: Read them. Every page. Read the footnotes. This may be the singular most important thing you can do to avoid unnecessary problems.

Design Analysis Newsletter, quarterly, Design/Analysis Consultants, Inc., Tampa, FL, (813) 265-8331

Troubleshooting Analog Circuits, Robert A. Pease, Butterworth Heinemann

Here's a book full of highly useful tips and observations on testing (and designing) analog circuits, by one of our country's leading electronics experts. It's always been our opinion that a well-written engineering book is worth more than its weight in gold; at $25 and packed with important insights, *Troubleshooting Analog Circuits* is a bargain. Highly recommended. (Available from Robert A. Pease, 682 Miramar Ave., San Francisco, CA 94112. Price includes tax and shipping.)

Intuitive Operational Amplifiers, Thomas M. Frederiksen. To order, send $25.00 to Thomas M. Frederiksen, 24705 Spanish Oaks, Los Gatos, CA 95030 (408/353-9415). (Price includes tax and shipping.)

Advances in Switched-Mode Power Conversion, Cuk and Middlebrook, TESLAco, Pasadena, CA

Modern DC-to-DC Switchmode Power Converter Circuits, Severns and Bloom, Van Nostrand Reinhold

Bipolar and MOS Integrated Circuit Design, Alan B. Grebene, John Wiley & Sons, New York

Waveforms, Homer B. Tilton, Prentice-Hall

HIGH-SPEED DIGITAL

MECL System Design Handbook, William R. Blood, Jr., Motorola

EMI, LAYOUT, GROUNDING

Books by Don White / Interference Control Technologies, Inc., State Route 625, Gainesville, VA 22065. A particularly good reference is *EMI Control Methodology and Procedures.*

Noise Reduction Techniques in Electronic Systems, Henry W. Ott, John Wiley & Sons, New York

Instrumentation Fundamentals and Applications, Ralph Morrison, Wiley-Interscience, New York

RELIABILITY, RISK ASSESSMENT

Probabilistic Risk Assessment, Henley and Kumamoto, IEEE Press

Sneak Analysis Application Guidelines, (RADC-TR-82-179), Buratti and Godoy, Boeing Aerospace Co. for Rome Air Development Center, Griffis AFB, NY

CALCE News, newsletter of CALCE Electronic Packaging Research Center, University of Maryland, (301) 405-5323

Although the body of available statistical source material is generally aimed at yield optimization (design centering) rather than worst case analysis, with a little mental juggling some approaches can be modified for worst case purposes. Suggested references are *Statistical Design of Integrated Circuits*, A. J. Strojwas, editor, IEEE Press, and *Tolerance Design of Electronic Circuits*, Spence and Soin, Addison-Wesley.

Flying Buttresses, Entropy, and O-Rings, James L. Adams, Harvard University Press

To Engineer Is Human: The Role of Failure in Successful Design, Henry
Petroski, Random House, NY

BAD SCIENCE

Galileo's Revenge, Peter Huber, Basic Books

For managers responsible for product quality and concerned with
liability exposure, *Galileo's Revenge*, by Peter Huber will confirm
your worst fears about our crazy legal system. Mr. Huber makes a
very convincing argument that many of today's judges allow "junk
science" to be introduced in the courtroom, which affords fringe
theorists of the caliber of psychics and astrologers the same stature
in the eyes of the jury as real scientists and engineers. As Huber
demonstrates, this often results in guilty verdicts when the best
scientific evidence shouts "innocent!"

Huber's book offers numerous interesting and often amusing examples of junk science case histories, ranging from 15th century
witch-burning (because witches caused plagues and crop failures)
to the recent Audi lawsuits (because Audis were capable of "sudden acceleration" on their own, causing injury to the unwary driver).
Of course, the power of witches has been greatly exaggerated, although it took 200 years and 500,000 witch burnings before this
was realized. And the Audi's engineering was just fine; the real
problem was that some drivers stepped on the accelerator rather
than the brake pedal.

Huber also shows how the harmful effects of junk science often
extend well beyond the conviction of innocent defendants. For
example, the Audi owners who switched models as a result of the
adverse publicity probably wound up with less safe vehicles and
higher related injuries and deaths (Audi was ranked at the top of
the safety list at the time). On the brighter side, Mr. Huber points
out that some judges are starting to restrict or even reject junk science testimony. However, until this becomes uniform practice, it is

easy to understand why many innocent defendants are willing to make out of court settlements rather than stand firm on scientific principles, knowing that those principles are likely to be debased by junk science in the courtroom. Until sanity has been restored to our legal system, *Galileo's Revenge* underscores the necessity of ample liability insurance and good, well-documented quality assurance procedures.

Betrayers of the Truth, Broad and Wade, Simon & Schuster

Science: Good, Bad and Bogus, Martin Gardner, Prometheus Books

Flim-Flam!, James Randi, Prometheus Books

Skeptical Inquirer quarterly (800-634-1610)

Reason magazine (815-734-1102)

MISCELLANEOUS

Free reprints of "What's All This Taguchi Stuff, Anyhow?" and "What's All This Fuzzy Logic Stuff?" are available from Robert A. Pease, National Semiconductor, MS-D2597A, P.O. Box 58090, Santa Clara, CA 95052.

APPENDIX A

HOW TO SURVIVE AN ENGINEERING PROJECT

Since human factors often strongly influence, if not dominate, the design and validation process, it is only appropriate that they be at least mentioned in a Handbook dedicated to design excellence. So come with us for a brief tour of the engineering environment while we follow a project from conception to completion.

SCENARIO # 1:
THE NEW PROJECT HAS TECHNICAL CHALLENGES WHICH ARE NOT REFLECTED IN THE SCHEDULE OR BUDGET

When a contract is won, it is not uncommon for the project engineer to find that the budget and schedule are marginal, and sometimes inadequate. This unfortunately results in pressure to trim back on planned effort, with analyses often being a prime target. As should be clear by now, there is one thing that should not be sacrificed:

THE PROJECT ENGINEER'S RULE FOR
TRIMMING THE FAT:
Never Cut The WCA

If this rule is not followed, then the result is the Engineer's Lament: "We never have time to do it right, but we always have time to do it over."

SCENARIO # 2:
COPING WITH "NOT INVENTED HERE"

Although WCA is very cost effective, it is still a complex and time consuming task. Therefore, an important way to reduce project costs is to make use of existing design and analysis data.

This is a ticklish area. The "NIH" (not invented here) factor is the tendency to start a design from scratch rather than using or improving upon an older design. Perhaps because a design reflects the creative capabilities of its originator explains why some designers feel compelled to leave a personal imprint upon their work. Nonetheless, assuming that specifications can be met by using or modifying an existing design, the project engineer has an obligation to place creative and artistic goals in their proper subsidiary position relative to the primary technical and economic project objectives:

THE PROJECT ENGINEER'S CURE
FOR "NOT INVENTED HERE"
Require the use of proven designs wherever possible

NIH
Although to spec be true,
A copy just won't do!
If its performance is superior
But its construction is familiar
After suitable review
We'll decide: It won't do!
And fix our attention instead
On the idea within our head.
It may not work at all,
But it's original!
There is no design so dear
As the one invented here.

SCENARIO # 3:
THE OVERDESIGN SYNDROME

The Overdesign Syndrome is first cousin to Not Invented Here. Some engineers love to not only re-invent the wheel, but to make it rounder, faster, and bigger. Quick to show disdain for any "routine" task ("any junior engineer can do that"), they love to jump headlong into the "challenging" (i.e., fun) aspects of a design effort. Because their devotion to their work is so intense, and because they may spend substantial amounts of after-hours effort at their work, it is difficult for the project engineer to criticize their efforts. Nevertheless, their activities may not be helping the design team.

The first problem is that the effort expended may simply not be necessary. Why spend days, weeks, or even months on improving a design when the original version will do? The second problem is that by focusing only on the major design topics, a multitude of smaller but equally important design details are left unattended.

The cure for the overdesign syndrome is the same as for NIH...use the existing design/analysis database to the fullest extent possible.

LET'S MAKE IT BETTER!

Project Engineer:	What's holding up the garbler design?
Designer:	Well, it's not easy to design an 18-point garbler.
PE:	But we only need a 12-point garbler.
D:	Yeah, but we can do a lot better.
PE:	But we only need a 12-point...
D:	And I'm almost there. This will be a real breakthrough!
PE:	But we only need a 12-point...
D:	Anyone can design a 12-pointer! They're available off-the-shelf.
PE:	Maybe you ought to order one.
D:	But in another couple of weeks I'll have you an 18-pointer!

HOMICIDE OCCURS AT XYZ COMPANY

UPI (Tampa) Police investigating the homicide of design engineer J. Jones at XYZ Company have obtained a confession from XYZ Project Engineer J. Smith. Other than admitting to the homicide, Smith does not appear to be very coherent, and mumbles continuously about "good off-the-shelf 12-point garblers..."

SCENARIO # 4:
THE SUPERVISOR'S SPOTLIGHT

It sometimes seems as though a well-conceived, thoroughly analyzed design which coasts into production with only minor modifications is treated as though the task was therefore routine, minor, and unchallenging. It thus follows that the task was executed by an engineer with unexceptional talents.

On the other hand, if the same engineer should have the misfortune to run into some difficulty, this all too frequently results in the entire attention of the corporate hierarchy being spotlighted on the errant design, accompanied by scowls and insinuations of the designer's inferior capability.

Those "unexceptional" designers who for the most part only turn out designs that work properly, on time and in budget, can be understandably irritated by a system which only shines the spotlight their way when they have their fair share of problems.

THE PROJECT ENGINEER'S SPOTLIGHT RULE
If the spotlight only shines when mistakes are made,
it will eventually shine on an empty stage

SCENARIO # 5:
A PENNY FOR YOUR HYPOTHESIS

When design problems persist, they create an increasingly stressful situation. Project engineers sometimes react to the stress in a haphazard fashion which tends to kill the one hope they have of putting things right: an organized and calm investigative effort.

The crisis atmosphere which begins to envelop a troubled program generates a wide variety of responses from the project team. In general, everyone wants to help, but everyone has their own notion of what "help" is. One common problem is that meetings often deteriorate into forums for unbounded hypotheses of what constitutes "the real problem" with the design. This cacophony reaffirms the old adage that opinions are a dime a dozen.

If the project is to survive a serious problem, the project engineer must insist on an objective investigation, supported by test and analysis data which have been subjected to peer review.

THE PROJECT ENGINEER'S GUIDE FOR SEPARATING
THE WHEAT FROM THE CHAFF
Don't accept anyone's opinion

Insist upon researched and documented support for technical hypotheses. This is extremely important, for it is the only valid way for a supervisor to evaluate technical inputs. Following this guideline will help to prevent the enormous waste of resources which otherwise occurs by pursuing every popular "quick fix" idea.

Be cautious of the engineer who is offended that their opinion is not accepted at face value, or who uses an excess of huffing and puffing to "prove" a point. Engineers who know their business will generally be happy to provide thorough justification for their recommendations, and will do so with a minimum of table thumping. Seek these out for technical advice.

SCENARIO # 6:
COPING WITH DESIGN PANIC

When the designers are stumped is when the project engineer must take control. If control is lacking, the trouble really starts, with the designers immersed in a high-pressure frenetic atmosphere that will make it a least three times harder to do their job. Lack of control will leave the design team faced with numerous interruptions for meetings, reports to upper management, enforced "brainstorming" sessions, and constant redirection of their problem-solving focus; i.e., a haphazard and confusing environment which reduces the chance for success.

The project engineer at this time should provide calm leadership:

THE PROJECT ENGINEER'S GUIDE TO CRISIS MANAGEMENT:
STEP 1
Prepare a problem solving plan

Using the best technical inputs available, including outside help if necessary, prepare a problem solving plan which is based upon a priority listing of activities. The plan should be informal and adaptive, i.e., easily modifiable to accommodate fresh evidence. However, a hypothesis selected for testing must not be capriciously abandoned, but must be pursued until sufficient results have been obtained. This is the only way to avoid the "fire-drill" syndrome, where activities are started based on the latest hot idea, and terminated without conclusions or documentation whenever the next inspirational flash occurs. At the end of the drill, the problem will probably still be unsolved, with the solution being trampled underfoot by the scurrying of the mob.

THE PROJECT ENGINEER'S GUIDE TO CRISIS MANAGEMENT:
STEP 2
Follow the plan

Using the plan, implement each activity. It is important to establish a calm atmosphere, since tension is easily communicated by body language and tone of voice. Terminate each activity only when valid results have been obtained. Document the results of each activity. Do not interrupt the designers with excessive meetings, report writing requirements, or other distractions. Implement an overtime schedule but don't overdo it; a fatigued designer is a poor problem solver. Promote the team atmosphere; don't leave any designer completely on the hook for a design that's in trouble.

THE PROJECT ENGINEER'S GUIDE TO CRISIS MANAGEMENT:
STEP 3
Don't accept unsubstantiated "solutions"

The problem solving approach should involve a blend of analysis and test activities. Beware of the designer constantly in the lab fiddling with the prototype, or at the computer terminal fiddling with a simulation. Although some tinkering is required to obtain an empirical feel for actual system performance, constant tinkering is a sign of desperation, fatigue, or both. Tinkerers may actually make things worse; by not really understanding the effects of their actions, they may appear to correct a problem when they have only patched the symptom, creating even more serious problems in the rest of the design. (This is called "squeezing the balloon.")

SCENARIO # 7:
ACHIEVING SUCCESS WITHOUT BURNOUT

The arduous effort required to achieve success on a high-intensity development project can generate a form of mental and physical exhaustion referred to as "burnout." There appear to be two misconceptions about burnout. The first is the macho mythology that human beings can be exposed to prolonged periods of intense stress with only minor ill effects. The second misconception is that burnout is a short term state which can be cured by taking the weekend off.

Burnout occurs over a long period of time, usually months. Its development is very difficult to detect, because its accompanying stress symptoms are ironically masked initially by the excitement of the activity that is creating the stress. After the body's reserves are depleted, however, when the stimulus of the creative challenge and the sixth cup of coffee can no longer lift the spirits, the crash occurs. The resulting fatigue, depression, ill temper and irrationality are not pleasant to experience or behold. Recovery times can vary, depending on the life style of the individual, but a few months of below par performance should not be considered unusual.

Most technical managers know that a stimulating environment helps productivity. However, there seems to be a lack of sensitivity to the signs which indicate that a project is entering the burnout zone. Once this zone is entered, stress becomes a negative factor, resulting in increasingly diminishing returns.

To help prevent burnout, it is imperative that "Theory X" management be avoided, wherein various forms of pressure and intimidation are used to "motivate" the design team to work long and hard despite burnout symptoms. Although Theory X management may on occasion achieve short term results, in general it is a proven recipe for disaster. Highly capable people cannot achieve unrealistic goals, no matter what pressure is applied. A work-each-night and work-each-weekend environment generates a proliferation of fatigue-induced errors, resulting in negative progress.

When a project is behind schedule and the design team has worked to its limits, there are often only two hard choices for the project engineer: to be late, or to be later. The application of Theory X management at this critical time will help guarantee the latter. On the other hand, a project engineer armed with a realistic appraisal of human capabilities and needs can squeeze as much success as possible out of a troubled project, and may achieve impressive or even spectacular results:

THE PROJECT ENGINEER'S GUIDE TO PROJECT SUCCESS
Priority 1: The Project Team
Priority 2: The Project Goals

APPENDIX B

THE DACI WORST CASE ANALYSIS STANDARD

1. SCOPE

This engineering standard defines the requirements for design validation by analysis, using enhanced Worst Case Analysis *Plus* (WCA+™) methodology.

2. SUPPORTING DOCUMENTS

The *Design Analysis Handbook*, DACI

3. REQUIREMENTS

Equipment design and design changes shall be validated by Worst Case Analysis *Plus* methodology (WCA+), which shall consist of a Functional Margins Analysis (FMA), a Stress Margins Analysis (SMA), and an Applications Analysis.

Problems identified by the analysis (ALERTs) shall receive a risk assessment based upon their calculated probability of occurrence. Recommended corrective action shall be provided where warranted by the risk assessment. For ALERTs requiring corrective action, sensitivities shall be computed to facilitate design corrections and optimization.

The analysis results shall be documented in a report. The report shall contain sufficient data to allow independent validation of the results.

3.1 FUNCTIONAL MARGINS ANALYSIS

A Functional Margins Analysis (FMA) is the calculation of the minimum and maximum values for each specified system function. The min/max values are compared to the corresponding allowable limits as defined by the system specification. Values which exceed specification limits indicate unacceptable performance (ALERT category); those which are within specification limits indicate acceptable performance (OK category).

Appendix B

The Functional Margins Analysis identifies those portions of the design where system performance exceeds specification limits; i.e., a "tolerance" analysis. Functional parameters and limits are usually defined by the equipment specification. Examples of functional parameters are gain, bandwidth, acceleration, flow rate, dimension, position, etc.

3.2 STRESS MARGINS ANALYSIS

A Stress Margins Analysis (SMA) is the calculation of the maximum applied stress for each stress parameter for each component in a system. The maximum stress value is compared to the corresponding allowable stress limit as defined by the component specification. Stress values which exceed stress limits indicate unacceptable overstress (ALERT); those which are within stress limits indicate acceptable stress (OK).

The Stress Margins Analysis identifies those portions of the design where absolute maximum limits are exceeded; i.e., a "breakage" analysis. Stress parameters and limits are usually defined under the heading "Absolute Maximum Ratings" on component data sheets. Examples of stress parameters are voltage, pressure, power, shear, intensity, temperature, etc.

3.3 APPLICATIONS ANALYSIS

An Applications Analysis is an evaluation of all applications, characterization, special test, and other miscellaneous and supplementary data related to the components and subassemblies employed in a design. Application violations are identified as ALERT cases.

The Applications Analysis is a "catch-all" supplementary analysis which is used to identify important design constraints which are not documented in the primary specifications. For example, a footnote at the bottom of an applications note may contain this important restriction: "A load capacitance exceeding 50 picofarads may cause oscillations."

3.4 SUPPORTING DOCUMENTATION

All calculations shall be based on data provided explicitly within the body of the analysis, or in an appendix to the report. Any interpolations, extrapolations, or assumptions used in deriving results shall likewise be presented with the related computations, or in an appendix. Supporting documentation shall include or reference as applicable:

- Specifications
- Schematics and drawings
- Parts Lists
- Derating Criteria
- Vendor specifications (data sheets) for components and subassemblies
- Vendor applications data
- Characterization data
- Miscellaneous data (special reports, etc.)

3.5 WCA REPORT OUTLINE

A report shall be generated which documents the results of the Worst Case Analysis. The report shall employ the following outline:

Title Page, including proprietary notice

Table of Contents

1. INTRODUCTION and SUMMARY

2. APPLICABLE DOCUMENTS

3. APPROACH

Description of computational techniques, algorithms, assumptions, and limitations. Definition of ALERT, CAUTION, and OK status categories.

4. ANALYSIS

Equations, variable values, calculations, and results for each analysis case (can be summary format with details provided in attachments).

Appendix B

5. RESULTS SUMMARY

Summary of ALERTs and CAUTIONs. Risk assessment and recommended corrective action for all ALERTs.

ATTACHMENTS AS REQUIRED:

Detailed calculation worksheets, data sources for equations and variable values, etc.

APPENDIX C

WORST CASE ANALYSIS SAMPLE REPORT

+24V REGULATOR

WORST CASE ANALYSIS

INITIAL RELEASE

28 June 1994

Prepared For

Mr. J. Smith, Technical Director
XYZ Corporation
1234 Business Blvd.
Tampa, FL, 33618

PROPRIETARY NOTICE

The data contained herein are the property of XYZ Corporation and may
not be disclosed to any other party without permission.

Appendix C

TABLE OF CONTENTS

1. INTRODUCTION and SUMMARY

2. APPLICABLE DOCUMENTS

3. APPROACH

4. ANALYSIS RESULTS

Attachments: Analysis Worksheets

1.0   INTRODUCTION AND SUMMARY

This report documents the results of the Worst Case Analysis of the +24V Regulator. The analysis shows the range of performance and applied stresses to be acceptable for operational reliability, excepted as noted below:

FUNCTIONAL MARGIN POTENTIAL PROBLEMS

a. The output of +24V Regulator U1 can exceed specification.
b. +24V Regulator U1 can have insufficient headroom.

STRESS MARGIN POTENTIAL PROBLEMS

a. Bridge Rectifier BR1 can have excessive junction temperature.
b. Filter Capacitor C1 can have excessive ripple current.
c. +24V Regulator U1 can have excessive junction temperature.
d. +24V Regulator U1 can have excessive input-to-output voltage during short-circuit conditions.

It is recommended that the analysis results of this Initial Release be compared to test data to validate the conclusions provided herein. Test/analysis discrepancies, if any, will then be reconciled and documented in updates to this analysis.

2.0   APPLICABLE DOCUMENTS

Schematic: 45376 Rev B, 10-22-93
Parts List: 45377 Rev C, 11-17-93

3.0   APPROACH

The Worst Case Analysis evaluates the range of functional performance and component stresses that will be experienced as a result of variations in component and operational tolerances. The analysis assumes that all components are in good working order and that they have been assembled correctly (no component or workmanship defects). The results of the analysis allow the design manager to evaluate the quality of a given design, and provide objective data by which potential design deficiencies can be evaluated for consequence, error magnitude, and probability of occurrence.

The Worst Case Analysis is structured into three components: a Functional Margins Analysis (FMA), which evaluates functional operation versus performance specifications; a Stress Margins Analysis (SMA), which evaluates component stresses compared to their maximum ratings; and an Applications Analysis Review, which evaluates component usage versus vendor recommendations.

3.1 FUNCTIONAL MARGINS ANALYSIS

The circuitry is analyzed to determine the min/max range of functional responses to initial settings, component tolerances, aging, temperature, supply voltages, inputs, and/or loads.

Circuit min/max responses are compared to required responses, and the results are categorized as follows:

ALERT: The min or max response exceeds the specification limit with a probability greater than 0.001%.

CAUTION: The min or max response exceeds the specification limit with a probability less than 0.001%, OR

Insufficient data; cannot fully evaluate, OR

Unquantifiable area of concern.

OK: Acceptable functional margin.

3.2 STRESS MARGINS ANALYSIS

Each component is evaluated with respect to each of its stress parameter limits (absolute maximum ratings) as defined by the specifications for the component. Worst case application values are calculated for each stress parameter. The results of each calculation are categorized as follows:

ALERT: The worst case value exceeds the absolute maximum rating with a probability greater than 0.001%.

CAUTION: The worst case value exceeds the absolute maximum rating with with a probability less than 0.001%, OR

Insufficient data; cannot fully evaluate, OR

Unquantifiable area of concern.

OK: Acceptable stress margin.

3.3 APPLICATIONS ANALYSIS

Applications data are reviewed for each component and compared to circuit usage to check whether the design contains usage violations.

3.4 RISK ASSESSMENT

The presence of an ALERT or CAUTION condition does not indicate a circuit deficiency; these status labels are used only to identify cases which receive further review. Each such case is evaluated to determine the consequences, magnitude, and probability of any related out-of-spec condition. If warranted by the evaluation, corrective action recommendations are provided.

4.0 ANALYSIS RESULTS

Analysis results are summarized below. Calculation details are provided in the attached worksheet.

4.1 FUNCTIONAL MARGINS (Worksheet REG24V.DM)

24V REGULATOR FUNCTIONAL MARGINS SUMMARY

Ref	Part #	Desc.	Parameter	Min Limit	Actual	Max Limit	Status
U1	LM317K	24V reg	V24	22.8	21.3/26.4	25.2	Alert 6.4%
U1	LM317K	Ilim. marg.	U1cl	0	0.524	NA	OK
U1	LM317K	Dropout	V24io	3	-2.8/22.7	40	Alert 6.6%

Appendix C

4.1.1 Discussion of ALERT and CAUTION Cases

a. +24V Regulation

The +24V output can exceed both the lower and upper specification limits as shown in the distribution below. The output will be beyond the specification limits approximately 6.4% of the time.

24V OUTPUT DISTRIBUTION

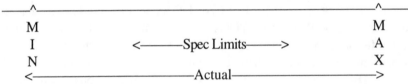

```
        |              XX            |
        |             XXXX           |
        L            XXXXXX          L
        I           XXXXXXXXX        I
        M          XXXXXXXXXXXX      M
        m         XXXXXXXXXXXXXXX    X
      XXXX | XXXXXXXXXXXXXXXXXXXXX|
   XXXXXXXXXXXXXXXXXX|XXXXXXXXXXXXXXXXXXXXXXXXXX|XXXXXXXXXXXX
____^_____^____
    M                                                     M
    I          <———Spec Limits———>                        A
    N                                                     X
    <————————————————Actual————————————————————>
```

The sensitivity analysis indicates that the error is primarily due to the initial tolerances of the LM317 regulator and resistors R1 and R2.

Corrective action options include:

• Use the "A" version of the regulator.

• Change the initial tolerance of the resistors to +/-1% or less.

• Use resistors with a smaller temperature coefficient.

• Add a snip or trimpot adjustment to allow calibration of the intial tolerance.

b. +24V Dropout

The LM317 requires a minimum of 3V in-out differential to ensure that its regulation specifications will be met. The analysis indicates that this differential, including input ripple, can go as low as -2.85V; i.e., the input can be lower than the output by this amount. During such conditions, the output will drop out of regulation. It is estimated that this condition will occur approximately 6.6% of the time. However, for low line voltages, this percentage will increase significantly (Vac sensitivity = 36%).

Corrective action options include:

*   Increase the size of the filter capacitor. A disadvantage is that the capacitor ripple current will increase (see 4.2.1.b).

*   Increase the transformer turns ratio and/or decrease the transformer output impedance. A disadvantage is that the maximum output voltage will increase (see 4.2.1.d).

4.2 STRESS MARGINS (Worksheet REG24V.DM)

24V REGULATOR STRESS MARGINS SUMMARY

Ref	Part #	Desc.	Parameter	Min Limit	Actual	Max Limit	Status
BR1	UT6	6A bridge	Vr		35/46	100	OK
BR1	UT6	6A bridge	Irms		.30/2.55	3	OK (1)
BR1	UT6	6A bridge	Isurge		24.8	65	OK
BR1	UT6	6A bridge	Tj C	-50	7.8/196	150	Alert 6.4%
C1	PA-1000	1000µF/50V	Volts		30.9/44.5	50	OK
C1	PA-1000	1000µF/50V	Irms		.27/2.05	1.3	Alert 75.7%
R1	CF2430	243 1% 1/4W	Watts		.006/.007	0.25	OK
R2	CF4321	4.32K 1% 1/4W	Watts		.097/.14	0.25	OK
T1	DVA-56-24	Xfmr, power	Watts		.15/11.17	11.2	OK
U1	LM317K	IC, +24V reg	Tj	0	4.5/219	125	Alert 14.8 %
U1	LM317K	IC, +24V reg	V24io		22.7	40	OK
U1	LM317K	IC, +24V reg	V24in		44.5	40	Caut (2)

Note 1: RMS ratings estimated at 50% of average rating.
Note 2: Max in-out volts with output discharge/shorted.

Appendix C

4.2.1 Discussion of ALERT and CAUTION Cases

a. *Bridge Rectifier Junction Temperature*

The junction temperature of bridge rectifier BR1 can reach 196 C compared to its maximum allowable limit of 150 C. It is estimated that this can occur 6.4% of the time.

Corrective action options include:

- Add a heat sink to BR1.

- Change BR1 to a type with a lower junction-air thermal resistance.

b. *Filter Capacitor Ripple Current*

The rms ripple current of filter capacitor C1 can reach 2.05 amps compared to the maximum rating of 1.3 amps. It is estimated that this can occur 76% of the time.

Corrective action options include:

- Change C1 to a type with an adequate ripple current rating. Since C1 must be changed, increasing its value can also reduce or eliminate the +24V dropout problem (see 4.1.1.b), but the rms operating current will be increased and the effects on BR1, C1, and T1 should be evaluated by an update to this analysis.

c. *+24V Regulator Junction Temperature*

The junction temperature of regulator U1 can reach 219 C compared to its maximum allowable limit of 125 C. It is estimated that this can occur 14.8% of the time. Although the regulator contains thermal shutdown protection, operation will still be adversely affected by nuisance shutdowns a significant percent of the time.

Corrective action options include:

- Use a large heat sink. Its sink-air thermal resistance should not exceed 0.131 C/W (see record #15 in worksheet).

- Use a somewhat larger heat sink coupled with the use of a higher-power/ lower thermal resistance regulator.

- Change the design approach for lower inherent dissipation (e.g., switchmode design).

d. +24V Regulator Peak Input Voltage

The peak input-to-output voltage across regulator U1 can reach 44.5 volts when the output is shorted compared to the in-out max rating of 40 volts.

Corrective action options include:

- Use a regulator with a higher breakdown rating, such as the LM317HVK, which has an in-out rating of 60 volts. Note: The LM317HVK will require the large heat sink described earlier.

INDEX

Index

Index

Newnes
An imprint of Butterworth-Heinemann

Related Titles

Introduction to EMC
by John Scott

John Scott's book provides a basic introduction to the principles of EMC. Essential reading, it supplies insight in what is needed to comply with the EMC Directive, and therefore opens the door to continued trading in Europe and the World.

January 1997 • 176pp • Paperback • 0-7506-3101-5

EMC for Product Designers
Second Edition
by Tim Williams

Widely regarded as the standard text on EMC, this book provides all the information necessary to meet the requirements of the EMC Directive. Most importantly, it shows how to incorporate EMC design principles into products, avoiding cost and performance penalties, meeting the needs of specific standards and resulting in a better overall product. This new edition includes the latest developments, which are essential for anyone complying to the regulations.

1996 • 312pp • Paperback • 0-7506-2466-3

http://www.bh.com/newnes

These books are available from all better book stores or in case of difficulty call:
1-800-366-2665 in the U.S. or +44 1865 310366 in Europe.